Recommendations for
Energy Efficient Exterior Lighting Systems

Published by The Institution of Engineering and Technology, London, United Kingdom

The Institution of Engineering and Technology is registered as a Charity in England & Wales (no. 211014) and Scotland (no. SC038698).

The Institution of Engineering and Technology
Michael Faraday House
Six Hills Way, Stevenage
Herts, SG1 2AY, United Kingdom
www.theiet.org

ISBN 978-1-84919-942-1 (paperback)
ISBN 978-1-84919-943-8 (electronic)

Contents

© The Institution of Engineering and Technology

List of Figures

List of Tables

Supporting organisations

The IET wishes to acknowledge the support received from the following organisations in the development of this Guide.

Department for Business, Innovation & Skills (BIS)

Highway Electrical Association (HEA)

Lighting Industry Association (LIA)

Scottish Futures Trust (SFT)

Zero Waste Scotland (ZWS)

Participants in the Exterior Lighting Systems Working Group

The IET would like to thank the following parties for their contributions to this Guide.

Lead technical contributors:
Phil Beveridge (4way Consulting)
David Dunn BSc IEng MILP MIET (Mouchel Middle East)
Steve Fotios CEng MEI MSLL MILE PhD, BEng(Hons) (The University of Sheffield)
Dave Franks (Westminster City Council)
Kevin J Grant IALD CEng MILP MIET MSLL (Light Alliance)
Allan Howard BEng(Hons) CEng FILP FSLL (WSP Parsons Brinckerhoff)
Paul Littlefair MA PhD CEng MCIBSE MSLL (Building Research Establishment)
David Lodge MEng MBA CEng MICE CPEng MIAust RPEQ (CU Phosco Lighting)
Nigel Monaghan MSLL (ASD Lighting/Institution of Lighting Professionals)
John O'Hagan BSc PhD CSci CPhys CRadP MInstP MIPEM MSRP MIES FLIA (Public Health England)
Luke Price MSci (Public Health England)
Gareth Pritchard BTech(Hons) CEng FILP MIET TechIOSH (Highway Electrical Association)
Paul Ruffles BSc MSLL (LD&T/Society of Lights and Lighting)

Additional contributors, participants and corresponding members:
Talia Addleman (Department of Energy and Climate Change)
Debbie Anderson IEng MIET CMHEA (Derbyshire County Council)
Eur Ing Matthew Clarke BEng CEng MBA MIET MCMI (Atkins)
Alan Hitch IEng MILP (Cambridgeshire County Council)
Tony Howells BEng (Department for Business, Innovation & Skills)
Julie Humphries (Lighting Industry Association)
Peter Hunt BSc(Hons) MIoD FIAM (Lighting Industry Association)
Lindsay McGregor BSc(Hons) CEng MIET MILP (Scottish Futures Trust)
Craig Mellis BEng (Hons) MIET (Salix Finance)
Eur Ing Ben Papé CEng FHKIE FIET FIMMM (on behalf of the IET) Chair
Sophie Parry AMSLL (Zumtobel Group)
Filipe Pereira-Lopes MILP MIET (Atkins)
Bernard Pratley (Lighting Industry Association)
Francis Pearce (Editorial consultant)
Russell Pryce AMILP (Balfour Beatty Living Places)
Paul Sargent BEng(Hons) MIET (Lighting Industry Association)
Mike Simpson BSc(Hons) FREng CEng FCIBSE FSLL FILP FIET (Philips Lighting)
Kelly Smith MSc BEng(Hons) AMILP AMSLL (Thorn Lighting)
Paul Smyth BEng(Hons) CEng MIET (Salix Finance)
David Taylor FIHE MIET (Transport for London)
Martin Valentine MSLL PLDA (Municipality of Abu Dhabi City)
Mayra Vivo-Torres (Department of Energy and Climate Change)
John Waite MSc CEng MCBSE, MSLL MIESNA (Arup)
Iain Watson BA(Hons) CA (Green Investment Bank)
Professor Arnold J Wilkins BSc DPhil FBPsS CPsychol HonFCO (University of Essex)

Secretary: Ian Borthwick BA(Hons) (IET)

Scope and purpose

Aim of this Guide

The aim of this Guide is to support customers in making informed decisions when acquiring exterior lighting systems. To achieve this aim, this Guide provides good practice in defining and delivering high quality, compliant systems that meet functional and budgetary requirements. It is published alongside a Good Practice Specification Template that provides a basis for specification of exterior lighting systems and the evaluation of supplier offers for suitable solutions.

Application

This Guide targets the application of lighting in locations in which new technologies offer improvements in energy efficiency, product lifecycle management, and adaptive lighting control. Such locations include public spaces, for example, outdoor work areas, road, footway and cycle path lighting, and security, architectural and amenity lighting.

Target audience

The target audience for this Guide includes those responsible for managing the procurement, specification and implementation of exterior lighting systems in both the private and public sectors, for example:

- technical, environment, transport or finance managers in local authorities,
- facilities or estate managers and property developers,
- exterior lighting specialists in manufacturing and supply companies, and
- lighting designers, lighting managers, consultants and contractors.

Introduction to energy efficient exterior lighting systems

1.1 Role of exterior lighting

The role of exterior lighting is to increase the utility and safety of outdoor spaces and to enhance the night time environment for its users. The overriding principle is that lighting should be 'human-centric' i.e. designed to deliver the required task, primarily for people. The decision to light an area and the choice of lighting technology depends on a wide variety of factors, not all of which are technical. Safety, comfort and enjoyment are the key criteria but care should also be taken to consider economic, ecological and heritage impacts of lighting within local settings.

Issues such as compliance, sustainability, value and aesthetics are essential to good lighting design and must therefore be integral to any exterior lighting scheme or masterplan. The lighting designer's role is to put the right light, in the right place at the right time (and in some cases this could be less light or none at all). The lighting technology used to achieve these aims is merely a means to an end.

▼ **Figure 1.1** Kings Cross Square lighting scheme (courtesy of Studio Fractal)

1.2 Decision-making and opportunities for exterior lighting

The simple fact that society operates by day and by night means that lighting affects almost every element of planning, from the economy to the environment. While safety is the top priority, the likely reaction of a community to a proposed lighting scheme is an important part of its design. For example, studies over the past two decades, such as those in the Cochrane Database of Systematic Reviews, have shown that street lighting can make roads safer and cut the number of personal injury accidents (PIAs). The introduction of white light can be effective in this respect but consideration also has to be given to its further aesthetic effects and to other complex and sensitive issues such as obtrusive lighting as well as light spill into residential areas and premises (in relation to intrusive light and lighting health concerns). The impact of a lighting scheme on flora and fauna is another important consideration, particularly in rural areas of outstanding natural beauty, where wildlife concerns, such as bat protection, may arise, as well as for sky glow (see 7.2 The Initial Design Process).

Lighting is a significant factor in economic regeneration and can help to increase social pride and cohesion. A well-designed city centre lighting scheme can help build the night-time economy, illuminating commercial and public spaces, façades and points of interest; enabling motorists to find their way and subtly directing foot traffic, as well as creating visual interest. The current understanding is that an appropriate lighting scheme, potentially combined with CCTV, can also reduce the fear of crime. Lighting's impact on actual crime is more complex, however, and any review of this subject must be evidence based. Advice on the role of exterior lighting in helping to prevent crime is provided in the Institution of Lighting Professional (ILP) publication *Lighting against Crime*, which is linked to a police initiative encompassing urban design in the UK, called Secured by Design.

A new lighting scheme should not be regarded as either a 'once and for all' solution or a one-off activity. Over time, schemes may need relighting to reflect changes in use. For example, with the trend towards a 24-hour economy, many roads and public spaces are used for extended periods of time. There may be more cyclists and pedestrians using the road than before and their needs differ from those of motorists. Rolling out the replacement of old lighting stock with more efficient, sustainable and cost-effective technology may solve a number of other issues. For example, installing LED lighting in preference to other light sources, can avoid the problem of mercury content in lamps while offering viable lighting solutions with reduced maintenance and lifecycle cost benefits.

Part of the lighting designer's role is to balance the cost and benefits of installing and maintaining lighting and its associated electrical network over the whole life of the lighting system. This assessment depends upon the installation design life, both in terms of the savings made against the existing lighting scheme's lifecycle costs, and any potential for the new system to generate income. A combination of improved lighting equipment and adaptive lighting controls offers immediate potential for enhancing public spaces at night while also saving energy and reducing maintenance costs. But a new lighting scheme could also provide a source of revenue, if the scope of works and the business case support deployment of additional technology. Planning for the future for lighting installations includes not only enabling the provision of energy-efficient components and fixtures as they become affordable, but also the hosting data and communications systems (or similar). In a growing number of towns and 'Smart Cities', lighting fixtures already share columns and infrastructure with technologies such as WiFi, CCTV, air quality and other monitoring systems, or electric vehicle charging points. (See the EU FP7 *Humble Lamppost* project for further detail).

▼ **Figure 1.2** Managing risks in the procurement of exterior lighting systems

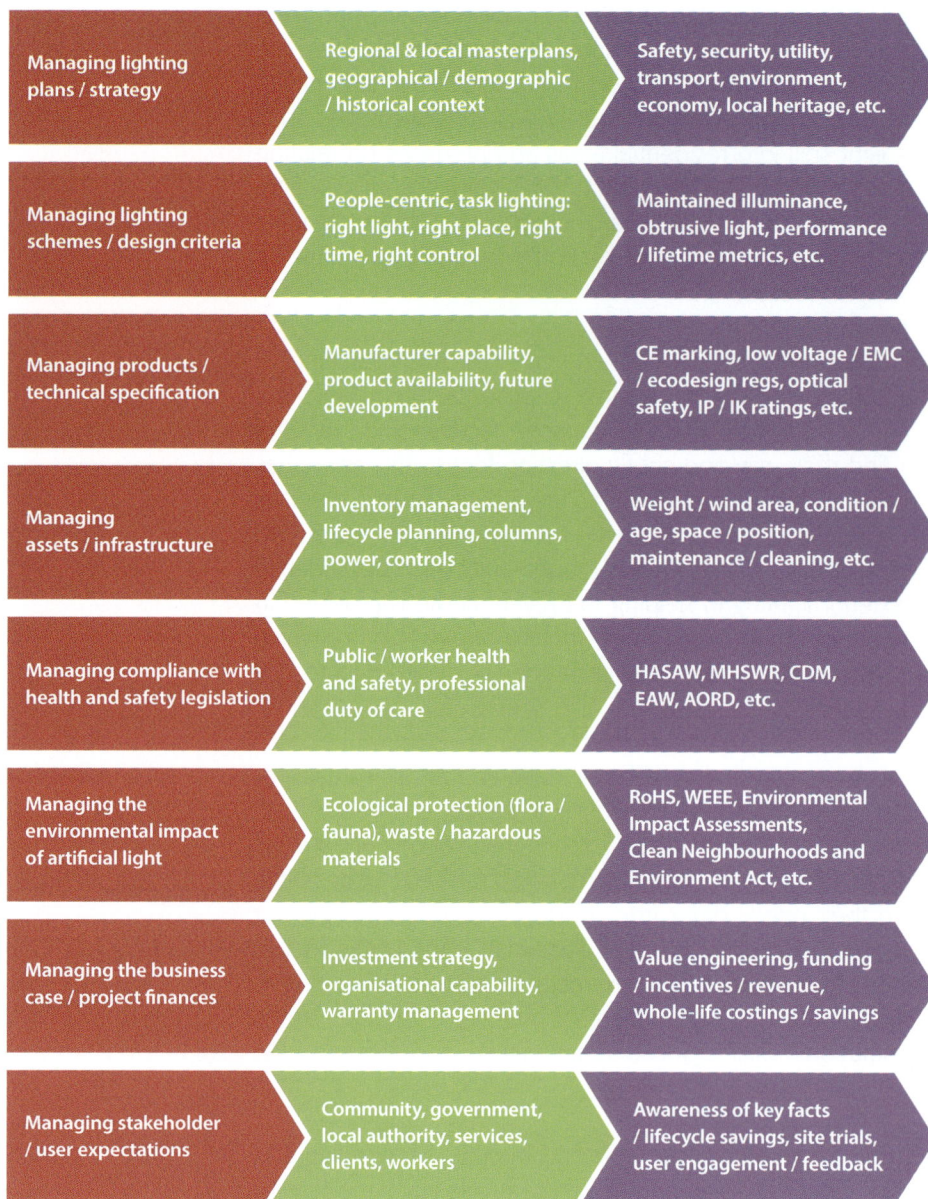

Managing lighting plans / strategy	Regional & local masterplans, geographical / demographic / historical context	Safety, security, utility, transport, environment, economy, local heritage, etc.
Managing lighting schemes / design criteria	People-centric, task lighting: right light, right place, right time, right control	Maintained illuminance, obtrusive light, performance / lifetime metrics, etc.
Managing products / technical specification	Manufacturer capability, product availability, future development	CE marking, low voltage / EMC / ecodesign regs, optical safety, IP / IK ratings, etc.
Managing assets / infrastructure	Inventory management, lifecycle planning, columns, power, controls	Weight / wind area, condition / age, space / position, maintenance / cleaning, etc.
Managing compliance with health and safety legislation	Public / worker health and safety, professional duty of care	HASAW, MHSWR, CDM, EAW, AORD, etc.
Managing the environmental impact of artificial light	Ecological protection (flora / fauna), waste / hazardous materials	RoHS, WEEE, Environmental Impact Assessments, Clean Neighbourhoods and Environment Act, etc.
Managing the business case / project finances	Investment strategy, organisational capability, warranty management	Value engineering, funding / incentives / revenue, whole-life costings / savings
Managing stakeholder / user expectations	Community, government, local authority, services, clients, workers	Awareness of key facts / lifecycle savings, site trials, user engagement / feedback

1.3 Using the Good Practice Specification Template

The Good Practice Specification Template provided in Annex A, and available to download (www.theiet.org/exterior-lighting) as an Excel spreadsheet, is designed to be used to set out and assess criteria so that alternative offers and lighting solutions can be assessed on a like-for-like basis.

The template has been designed to be broadly an output specification in terms of lighting equipment, and not to prevent the use of products that are part of a fast developing technology through the use of overly prescriptive input requirements. What is important is that the right type and amount of light is available for the required period of time (the lighting design life of the installation) and covers the required area in line with the appropriate standards.

NOTE: The specification template contains a significant number of parameters (aligned with relevant national and international standards) and so only occupationally competent individuals trained in the relevant IET Guidance should use and interpret this spreadsheet.

1.4 Capability and competence

Any client commissioning an exterior lighting project must comply with the requirements of the Construction Design and Management Regulations (CDM) 2015. Under CDM, duty-holders cannot appoint a Principal Designer, Designer, or Contractor unless they have taken reasonable steps to ensure that the organisation or individual they propose to appoint has the skills, knowledge and experience, and capability necessary to fulfil the role. This applies to any party involved within the design be it at the concept or feasibility stage through to detailed design, specification, construction and maintenance stages.

It is essential to ensure that manufacturers and organisations that manage exterior lighting systems are capable of so doing and that all designers, installers and maintainers are appropriately skilled and competent (particularly those acting in the capacity of the Principal Designer under CDM).

Professional competence in this respect can be broadly summarised as having sufficient knowledge, skill and experience – that is maintained or may be improved upon or broadened through Continuing Professional Development (CPD) – to enable an individual to:

- undertake the tasks required;
- identify the risks involved;
- recognise their limitations to make appropriate decisions based on sound engineering judgement; and
- take appropriate action to prevent harm to others who may be affected by the individual's acts or omissions.

All aspects of the lighting design and its implementation should be carried out by competent professionals. Clients should prepare their own competency matrices, based on their specific requirements. A variety of documents exist to assist in meeting the relevant competency, including Highways England GD 02, and ILP Competency Requirements for Lighting Design Staff.

All designers must be competent to identify and where practicable design out hazards and risks, including through the preparation of Design Risk Assessments (DRAs). DRAs should be prepared before and during the design process to remove risk to the contractor, or where not possible to remove at minimum provide mitigation to reduce any risk. The advice on the definition and responsibilities of a designer as they relate to CDM can be found on the Health and Safety Executive website (www.hse.gov.uk). Risk assessment can also include whole-life cost/benefit analyses of the lighting solutions proposed. A DRA may also be necessary to determine if any special requirements relate to a particular site.

1.5 Relationship between this guidance and the SFT/LP Street Lighting Toolkit

This Guide and the Good Practice Specification Template are intended to provide good practice guidance for the technical management of exterior lighting system specification, procurement and application. While high level coverage is provided for financial management and asset management in Sections 4 and 5 of this Guide (respectively), a more detailed approach to combined technical and financial modelling for exterior lighting is provided by the Street Lighting Toolkit published by the Scottish Futures Trust (SFT) and by Local Partnerships (LP).

The Street Lighting Toolkit was originally developed by the SFT as part of the Scottish Government's National Street Lighting Energy Efficiency Programme. LP, supported by the Department of Energy and Climate Change (DECC), are leading on the use of the Street Lighting Toolkit by local authorities in England.

The Street Lighting Toolkit assists in the identification and implementation of street lighting or exterior lighting projects that will improve energy efficiency, reduce carbon emissions and generate significant financial savings. It allows users to assess their energy efficiency investment needs based on the condition of their existing lighting stock and to extrapolate the potential whole life saving (financial, energy and carbon) from various programmed technical solutions.

The Street Lighting Toolkit and associated user guide are available free of charge via:

- the SFT website: www.scottishfuturestrust.org.uk
- the LP website: localpartnerships.org.uk/our-work/growth/efficient-lighting/

NOTE: The Street Lighting Toolkit is geared towards local authority highway lighting applications, in that it assumes unmetered connections and appropriate Elexon (UMSUG) codes from which to derive circuit watts. For other applications (for example, public buildings such as hospitals, schools, offices or commercial public spaces), it may be possible to insert dummy codes which match the actual circuit watts of the proposed lighting solution.

SECTION 2

Exterior lighting design

2.1 Introduction

This Section gives an introduction to the principles behind lighting design and adaptive lighting, and the standards and considerations that are applicable to a variety of exterior lighting applications. Sections 5 and 6 discuss the specification and design of exterior lighting systems in more depth.

NOTE: This Section is not intended as a comprehensive guide to lighting design and reference should therefore be made to other guidance documents published by the Institution of Engineering and Technology (IET), British Standards Institution (BSI), International Commission on Illumination (CIE), International Electrotechnical Commission (IEC), Society of Light and Lighting (SLL) and Institution of Lighting Professionals (ILP). Appropriate reference is required to key standards and guidance which are listed in Annex D.

2.2 Lighting design principles and terminology

2.2.1 Introduction to lighting design

Lighting design covers a range of considerations but its first priority is to provide the right lighting levels in the right place at the right time for tasks being lit to be carried out safely and effectively.

Being able to see in order to move around safely and carry out tasks requires enough light on the task itself but also limiting glare and ensuring visual comfort. The aim is to use light efficiently, putting the right light only where and when it is needed.

Exterior lighting design should seek to minimise obtrusive light where practicable, either to sensitive areas or buildings (receptors) nearby or upwards to the sky (sky glow). Due consideration for buildability, energy efficiency, maintenance and whole-life costs are part of the total lighting design process for all lighting installations.

The following points explain the key lighting terms that are relevant to the design of exterior lighting systems.

2.2.2 Maintained level (of average illuminance or luminance)

The amount of light available falling on a task or surface is given by the illuminance, defined as the luminous flux incident on a unit area, expressed in lux. Luminance is the brightness of a surface to an observer, defined as luminous intensity of light reflected from a surface measured in candela per square metre (cd/m^2).

Many lighting standards and codes use maintained level of illuminance or luminance as the key measure for lighting design. They contain ranges of recommendations for maintained levels for indoor and outdoor spaces.

Illuminance or luminance is called *maintained* as this is the design level that follows consideration of maintenance factors to allow for depreciation of the light source and luminaire performance. Typically, this refers to the lowest level of lighting during the design life of the lighting installation – i.e. at the end of life of the luminaire, when light sources have aged but have not been replaced and luminaires have deteriorated and cannot be cleaned back to their original condition.

Without some form of constant light output feature (see below), higher illuminances would normally be produced initially by the new lamps and clean luminaires to ensure that, given the depreciation in light output over time, the required light levels were still met at the end of the scheme life.

2.2.3 Light distribution

The light distribution of a light source may be represented by polar curves, which enable the visualisation of the luminous intensity at different angles in a number of vertical planes crossing the light source or by an isolux diagram showing the varying illumination over the lit surface from a single light source.

2.2.4 Beam angle and luminous intensity

The beam angle is an essential characteristic of directional lamps and luminaires, as it provides the width in degrees of the beam portion in which the luminous intensity of the light emitted is at least half the peak intensity that typically occurs at or around the centre of the beam. Beam angle is quantified in degrees and the (peak) luminous intensity is measured in candelas.

For example, if an installation is properly designed, floodlights with wide beam angles can light up large areas to a relatively uniform level of illuminance, in contrast to a spotlight that will have a narrow beam angle.

2.2.5 Directional lighting

Directional, as opposed to general, lighting is mainly used for task or accent lighting providing modelling, for example, for building façades, bridges and monuments. Care should be taken to provide sufficient light for the purpose while also avoiding high contrast with surrounding areas and/or light spill. (See uniformity, below). Good lighting design relies on a balanced mix of light, dark and shade, both for practical and aesthetic reasons.

2.2.6 Uniformity

While contrasting patterns of light and shadow can make a space look more appealing and animated, uniformity is important for work areas and street lighting to avoid dark areas where it is hard to see. BS EN 13201-2 *Road lighting Performance requirements* contains recommendations for minimum overall uniformity for illuminance designs, and for both minimum overall uniformity and longitudinal uniformity levels on the road, depending on the road lighting class, for luminance designs.

2.2.7 Obtrusive light

Efforts should be made to limit obtrusive light from outdoor lighting installations. Obtrusive light can intrude into homes, gardens or other outdoor areas and can also affect wildlife. Upward light from the installation causes sky glow.

▼ **Figure 2.1** Obtrusive light from a typical street lighting luminaire (image courtesy of BRE)

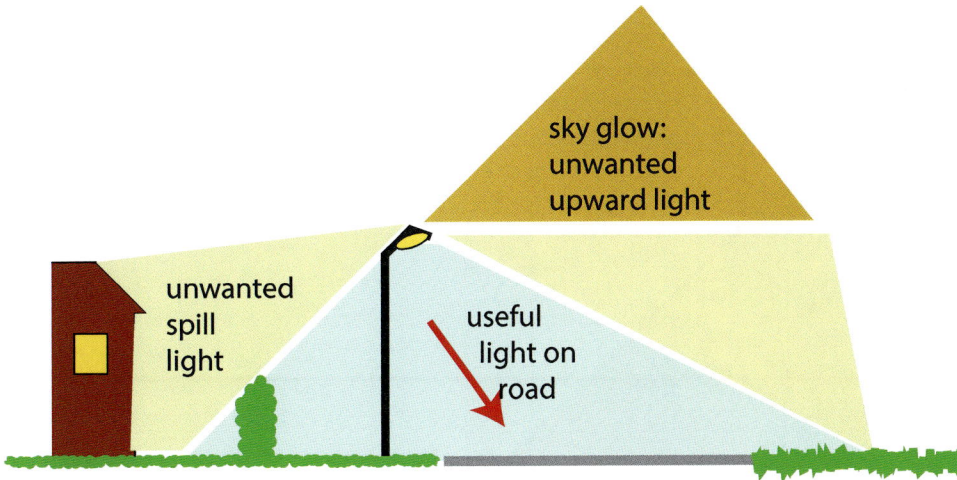

▼ **Figure 2.1** Obtrusive light from a typical street lighting luminaire (image courtesy of BRE)

Guidance on light pollution is given in BS EN 12464-2 *Light and lighting. Lighting of workplaces. Outdoor workplaces*, CIE Publication 150 *Guide on the limitation of the effects of obtrusive light from outdoor lighting installations*, CIE Publication 126 *Guidelines for minimising sky glow* and in ILP GN01 *Guidance Notes for the Reduction of Light Pollution*.

The limits in the guidance depend on the location of the site (for example, whether it is an urban or rural site). The CIE and ILP Guidance Notes also introduce the concept of a lighting curfew, where lighting is switched off or reduced at set times (the ILP suggests between 23.00 hours and dawn, where appropriate considering the task to be lit).

2.2.8 Glare

There are two types of glare: discomfort glare which makes one instinctively look away, and disability glare which makes it difficult to achieve a task. An example of the latter would be dazzle from headlamps, but disability glare can also be caused by poorly sited or angled artificial lighting.

Disability glare from lighting can be avoided by correct aiming of light towards areas of interest, as well as by the use of suitable shielding against direct view of lamps and high luminance parts of luminaires.

Glare and obtrusive light can be reduced by creating and orienting the right beam of light that is needed to light the specific external area.

2.2.9 Colour appearance

The colour appearance of a white light source is characterised by its colour temperature expressed in Kelvin (K).

The correlated colour temperature (CCT) of a light source is defined as the temperature at which its colour matches the colour of the radiation emitted by a heated ideal black-body radiator.

▼ **Figure 2.2** International Commission on Illumination (CIE) chromacity diagram (copyright Havells Sylvania)

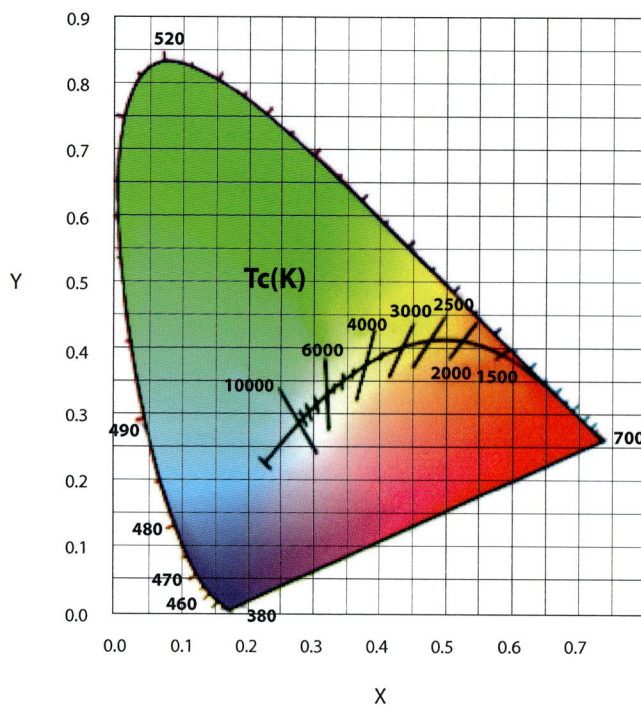

The correlated colour temperatures (CCT) of white light sources ranges from a warm near-yellow hue of white (2700K), through neutral whites (~4000K) to a much cooler bluish white (~6500K) and beyond.

▼ **Figure 2.3** Colour appearance chart

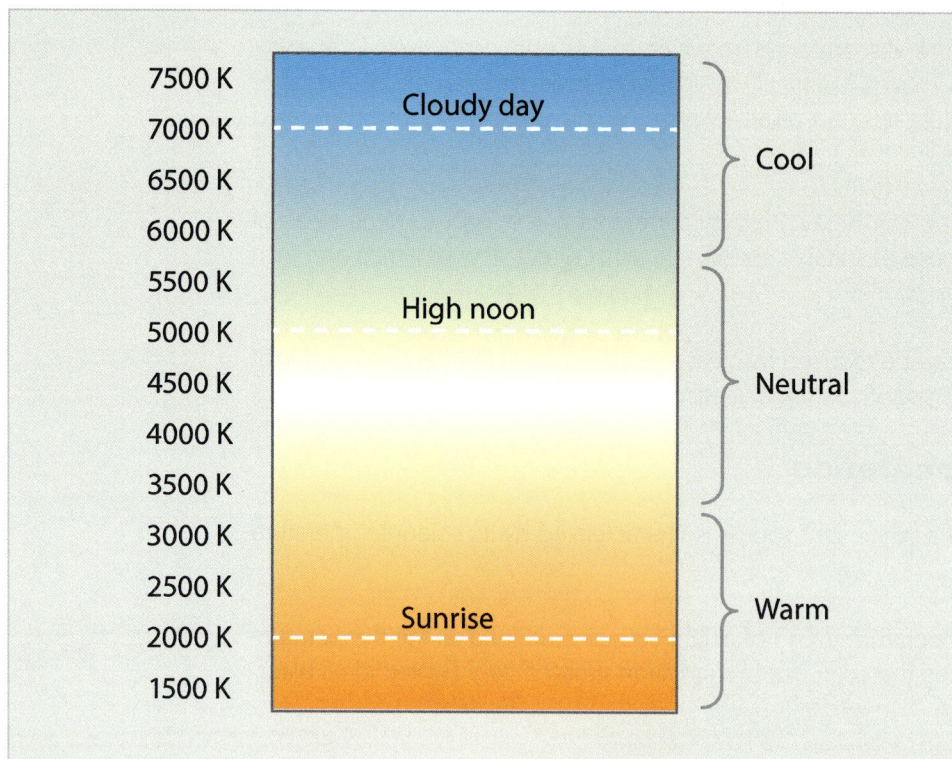

Although having the same colour temperature marking, lamps/light sources produced by different manufacturers may have slightly different colour appearance. To ensure colour consistency when installing or retrofitting lighting, lamps of the same colour temperature and preferably from the same manufacturer should be used.

Colour bins, or MacAdam ellipses on the chromaticity diagram (Figure 2.4), are used to specify LED lighting – most people would not see any difference in the colour of the light emitted falling within a 1-step MacAdam ellipse, and some colour difference starts to be noticed in the case of a 2-step MacAdam ellipse, which is currently considered as good practice.

▼ **Figure 2.4** Part of the CIE chromacity diagram showing the black body curve for white lamps of different colour temperatures. The small ellipses are two-step MacAdam ellipses. Lamps with chromacity coordinates within the same ellipse would appear to be the same, or almost the same, colour.

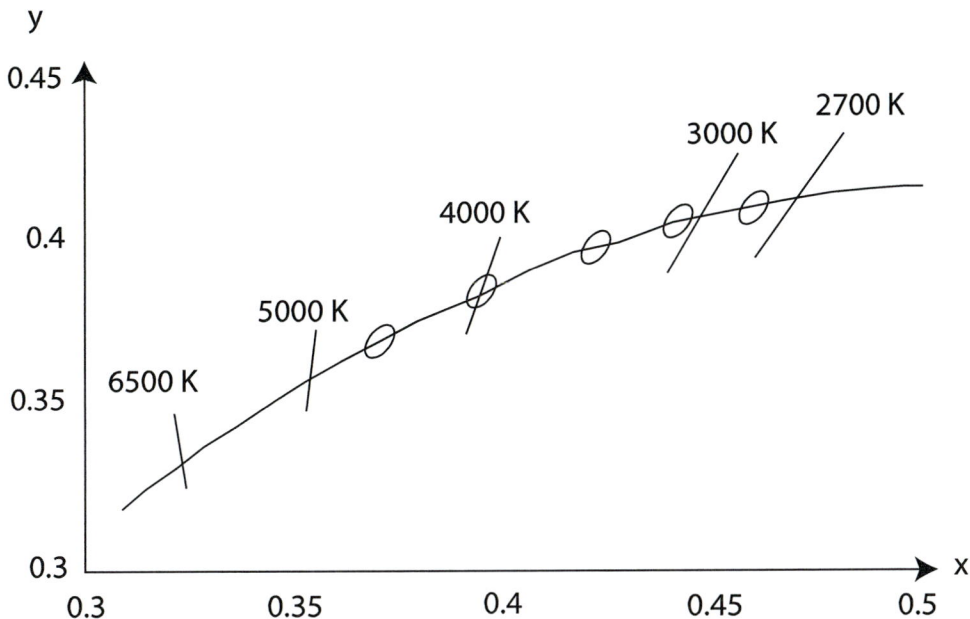

2.2.10 Colour temperature stability

Colour temperature stability refers a light source's ability to maintain its colour properties over time. Colour shift can have causes ranging from phosphors degrading within a lamp to the ageing of the materials used in its optics or of the driver. LED and other lamps – such as some high pressure discharge lamps – may have variations in colour appearance due to the manufacturing process. Some lamps may experience a colour shift over their lifetime.

Although less critical in exterior lighting applications than, say, museum lighting, colour shift leads to a loss of performance and inconsistencies in the appearance of lighting schemes. This can cause maintenance problems in that although lamps produced by different manufacturers may have the same colour temperature marking they may also have a slightly different colour appearance.

2.2.11 Colour rendering

Colour Rendering Index (CRI) indicates how closely colours of objects lit by an artificial light are shown compared to if they were lit by daylight. The higher the CRI, the closer the light source renders a designated range of colours to their appearance under a reference source. Colour rendering is expressed by the CIE general CRI with a value of 100 (the maximum) indicating a precise colour match and values less than 100 indicating differences in colour appearance.

The appropriate standards and guidance documents advise on the colour rendering depending on the task to be lit. For example BS 5489-1 *Road lighting* contains a recommendation for a CRI ≥ 60 in areas with high pedestrian use such as shopping streets and civic centres. Under the Ecodesign regulations, LEDs for exterior use should have a CRI of 65 or better.

2.2.12 Scotopic/Photopic (S/P) ratio

The units in which lighting is measured correspond to the response of the human eye in the day time; this is termed photopic vision. At night, for example, under dim moonlight, the vision is called scotopic vision and the spectral response of the eye shifts towards the blue end of the spectrum.

The ratio of the luminous output of a light source with reference to the ambient lighting levels and the CIE visual response of the human eye is termed the Scotopic/Photopic (S/P) ratio.

All light sources have an S/P ratio. Lighting of higher S/P ratio permits better visual performance under mesopic conditions (see below). If the S/P Ratio is equal to one the lamp performs equally under photopic, mesopic and scotopic conditions. A value greater than one indicates that the lamp produces more scotopic lumens than photopic lumens. A value less than one indicates that the lamp produces more photopic lumens than scotopic lumens.

Generally the whiter the light source, the higher the S/P ratio. When using lamps of CRI ≥ 60, BS 5489-1 introduced a reduction in design illuminance for subsidiary roads, with the reduction being greater when using lamps of higher S/P ratio. Depending on the efficiency of luminaires with the different S/P ratios, this lower lighting level may enable lower energy use, all other things being equal.

2.2.13 Mesopic (night time) vision

In between scotopic and photopic vision, i.e. with luminances between approximately 0.01 and 10 cd/m^2 or illuminance between 50 lux and 0.05 lux (which includes the range at which most exterior lighting operates), the response of the eye is called the mesopic vision range. Within this range, the eye becomes less sensitive (in relative terms) to yellow and red light, and more sensitive to shorter wavelength blue and green light (known as the Purkinje effect). Hence the orange light emitted by low-pressure sodium lamps may appear not to light the area as well as whiter light sources.

Due to the eye's improved response to whiter light in the mesopic range, lighting of higher blue content (for example, higher S/P ratio) can create an outdoor scene which may appear as bright as one created by other lamp types but at a lower ambient illuminances. Consequently, white light is most effective for areas with lower illuminances, rather than major roads where light levels will be higher. However there is also the colour appearance/colour temperature to be taken into account as many people do not like a 'cold' appearance.

2.2.14 Lifetime and overall maintenance factor

Lifetime of a luminaire is the point at which the luminaire fails to produce any light (not including normal lamp changes as part of maintenance) or where the light output falls below the maintained light output and so produces insufficient light to carry out the intended task safely.

The light levels recommended in this Guide and other guidance documents are maintained levels, at the end of luminaire life. Guidance on the calculation of maintenance factors is given within the appropriate standards.

The maintenance factor takes into account dirt on optical surfaces of the luminaires (both inside and out) and the decline in light output of the source with time due to depreciation of the source and, for LED luminaires, the failure of any individual LEDs within the LED module.

Constant light output controls (as described below) may be used to reduce the output of light sources during the early part of their lifetime, in order to avoid the over-provision of lighting and minimise energy use.

Manufacturers should declare luminaire life and overall maintenance factors at a specified ambient temperature.

2.3   Lighting controls and adaptive lighting

2.3.1   Lighting controls

There are a number of luminaire controls available in the market to enable lighting to be adapted to the task required within the exterior application. The following points explain key terms relevant to adaptive lighting for exterior lighting systems:

2.3.2   Constant Light Output (CLO)

CLO (sometimes called Maintenance Factor Harvesting) allows the lumen maintenance to be managed over time by increasing the drive current over the operating life of the luminaire to compensate for a gradual loss of light output based on assumed characteristics such as lumen depreciation.

In discharge lighting this is usually only offered as a benefit of a Central Management System (see below) but with LED lighting this facility can also be provided within the LED driver.

▼ **Figure 2.5** Typical maintained illuminance lifecycles for a range of exterior lighting systems

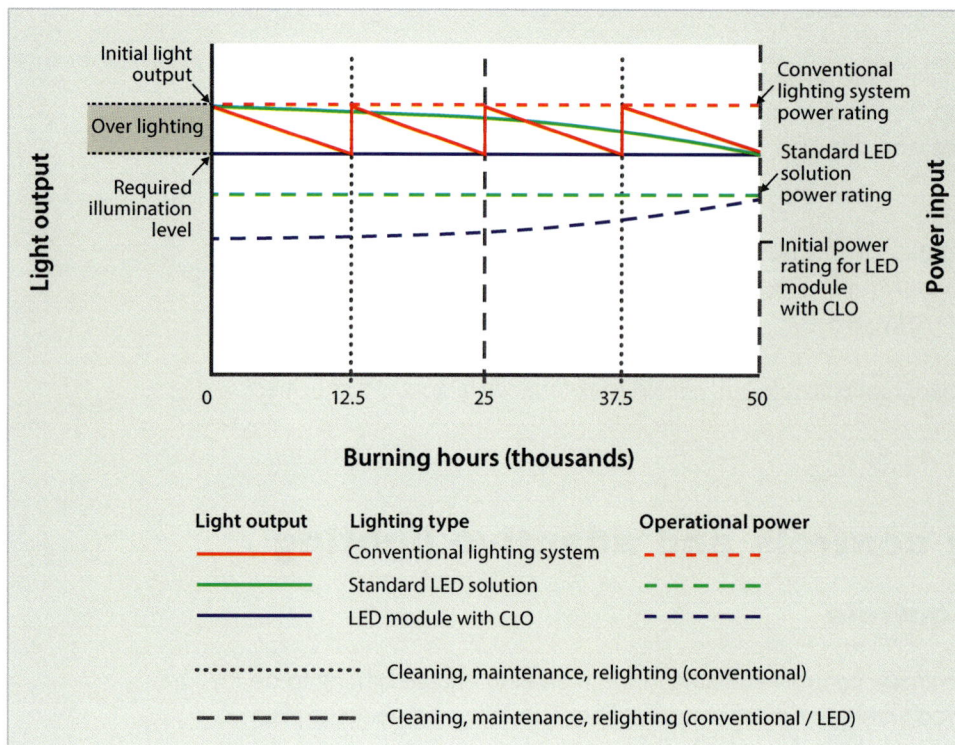

2.3.3 Adaptive lighting

Adaptive lighting is the ability to change the lighting levels based upon the task being lit. Adaptive lighting is a name for a group of controls that allow light outputs to be modified during the night and includes, Central Management Systems (CMS), part night lighting, stand alone dimming (single level static dimming using full power and a single fixed dim level/time window), and Multi Level Static Dimming (full power and multiple fixed dim levels/time windows), and dynamic dimming.

BS 5489 provides for adaptive lighting of a road where demand varies during the course of the night. Control technology fitted to luminaires enables adaptive lighting during periods of low usage, saving energy. See also ILP PLG08 *Adaptive lighting*.

1   CMS is usually a mixed wireless/GPRS communication system that links software user interface on the customer's PC or mobile device via the web-based link to a mainframe server. The service hosts the lighting control service and communicates with the network of luminaires, either directly or through a series of collector units, for example, via GPRS mobile phone data service to CMS nodes and then on to the individual luminaires via Zigbee or another wireless network protocol. The terminology for collectors and nodes varies with each different manufacturer but these are all essentially aggregating signals from lots of nodes and sending them back to the mainframe and, vice versa, distributing the mainframe instructions out to the nodes. The number of nodes that can be operated on a single collector varies from system to system from one or two hundred to several thousand nodes.

A CMS in itself does not provide, but rather, facilitates the energy savings that can be achieved through trimming, part night lighting and dimming by any other means. The additional benefits from CMS are centred around the real time control and reporting of luminaire operation and faults. CMS also provides a degree of future-proofing of the luminaires to be able to respond to changes in designed light levels in future

22

standards where these offer additional energy savings, or to respond to changes in the use of the area being lit. CMS where used with dynamic ballasts/drivers also offers a lighting engineer the opportunity to respond to resident's specific concerns about illuminance and luminance levels on a street-by-street basis.

2 Part-Night Lighting is where lighting is turned off during certain hours of the night, normally relating to low task usage (for example, residential roads between say midnight and 5.30 am) and can be applied via photo-electric control units (photocells/PECUs), suitable electronic control gear/LED drivers or by CMS.

3 Dimming is similar to part night lighting but rather than turning off the lighting, the luminaire light output is reduced to an acceptable level for the task (for example, roads where there are significantly reduced traffic flows outside peak times). The convention for specifying dimming levels uses 100 % dim as light/full power on and 0 % is switched off. Discharge lamps, control gear and LED drivers may have a minimum dim level limited by the equipment, where below this level the light switches off. Dimming can be achieved using discharge luminaires with electromagnetic control gear with twin tapped ballasts, electronic control gear with dimming function and dimmable lamps, or with LED luminaires with dimmable drivers.

4 Historically, only one step dimming tended to be used but recent technology changes allow more steps to be considered, to suit the task. This is referred to as Multi-Level Static Dimming (MLSD). 'Multi Level' indicates more than one level of dimming is used each night, while 'Static' means that once the regime is factory-set, the dimming levels/times may not be able to be changed on site.

MLSD systems allow a step change in lighting levels to be set at specific times of the night, by modifying the dimming level. Typically, systems can offer up to a six-stage timed cycle (five lighting levels and off). The dimming levels (down to 1 % interval) and times (to the nearest minute) are programmed into the driver or separate MLSD module during luminaire manufacture.

CMS can perform the same multi-level dimming functions but these are not truly 'static' as they are not factory-set.

MLSD devices are usually used for more complex dimming where CMS is not installed. MLSD functions may also provide switching functionality by setting a dim level of 0 % (i.e. OFF) during the selected period.

5 Dynamic dimming uses the active measurement of usage (for example, of traffic flows) to determine the design lighting class in real time and then adjusts the light levels to meet that class. At the time of writing, dynamic dimming is being trialled but is not widely available. It is likely to need a more complex control system and so is likely to be an optional feature within Central Management Systems.

In the cases of part night lighting, dimming and MLSD, the activation may be by timer or solar clock and relay, photocell (which needs analogue or DALI output signal for dimming), by control functions integral to the control gear/LED driver (which uses the photocell or timer switch on/off point to calculate approximate times to activate) or via an analogue or digital control signal from a CMS.

Once installed, adaptive lighting equipment (other than CMS) usually take a few days operating at 100 % to learn the dawn and dusk times before they calculate the start and end times for each dimming step in order to operate normally. The daily switching control device (timer, solar clock, photo-electric cell) is used to define the switch on and switch off times and the driver/MLSD device will calculate the intermediate dimming times. Some discrepancy in absolute timing may be observed due to seasonal changes, but particularly over the change between Standard and Daylight Saving Times.

2.3.4 Trimming

Trimming represents the reduction of the luminaire burning hours by tuning the ambient light level (in lux) that the luminaire switches on and off at dusk and dawn and applies to all light sources. Historically, in the UK, the switch on/off has been controlled by photocells set to 70 lux descending (switch on) and 35 lux rising (switch off), known as 70/35 lux photocells. The asymmetric levels are chosen to allow in the region of 15 minutes time for discharge lamps to warm up to full light output.

For LED and other sources that produce full light output instantaneously, this warm up period is not required to reach the required lumen output and a symmetrical switching can be used, for example 35/35 lux photocell, to give the same level of benefit to the lit area. Further savings can be made if the light level at switch on/off is reduced to suit the designed illumination levels. For example, an area lit to 20 lux average could in theory use a 20/20 lux photocell or, to allow a margin of safety, a 35/35 lux photocell. A minor road lit to 5 lux average could similarly use a 5/5 lux photocell or maybe an 18/18 lux photocell. This allows a few minutes of burning to be saved at the start and end of each night.

Together these two modifications to standard practice can offer energy savings from 2 % to 4 %.

2.3.5 Additional sensor and control systems

In some circumstances, it may be necessary to consider additional sensors, control technologies and/or combinations of both with lighting systems (for example, passive infrared for security lighting, or CCTV systems for highways). Where applied, these systems should be designed in line with the lighting scheme and comply with relevant planning regulations and standards.

2.4   Exterior lighting applications

2.4.1 Purpose of exterior lighting

The purpose of lighting the external environment is to make it easier, safer and more pleasant for people to use and to promote the economy of the area. This broad definition includes providing functional lighting to pedestrian areas and all classes of traffic route, but also covers areas such as signage, security lighting, access roads, car parks, docks and architectural lighting.

Typical exterior lighting applications include such public spaces as:

- outdoor work areas;
- road, footway and cycle path lighting;
- area lighting for security;
- sports floodlighting;
- external architectural and amenity lighting (including lighting for access to buildings, parks and public areas); and,
- illuminated signs.

2.4.2 Lighting of outdoor work areas

Guidance on lighting of outdoor work areas is given in EN 12464-2. This gives recommendations for maintained average illuminance, uniformity, glare and colour rendering in a wide range of different types of workplace. A wide range of light sources are available to the designer that can be useful for outdoor work areas as part of a combined task/background lighting approach. In addition, the overall lighting of an outdoor space may be supplemented by task lighting of specific areas. This approach can reduce energy consumption and through the use of luminaires with good optical control mounted correctly can also reduce obtrusive light. The lighting can be switched off or dimmed when the required tasks are not being undertaken, subject to such adaptive lighting being based on an appropriate risk assessment.

2.4.3 Road, footway and cycle path lighting

Lighting design of road, footway and cycle path lighting is carried out in accordance with BS EN 13201 and using the guidelines in BS 5489-1 *Road Lighting*. The latter document gives guidance for traffic routes, conflict areas such as road junctions, and subsidiary roads, paths, car parks and amenity areas. These are classified into categories depending on factors such as the road hierarchy and levels of traffic flow. BS EN 13201-2, CIE 115 and PD CEN/TR 13201-1 give recommended lighting parameters (for example luminance, illuminance, and glare) corresponding to each category.

It is important to ensure that roads are classified correctly to ensure the correct task lighting and energy use and the standards map to the UK road hierarchy gazetteer. BS 5489-1 requires the designer to follow a five stage risk based approach to the design process; the tables provide initial guidance to the lighting levels required which must then be considered following a site based risk assessment regarding the task to be lit and local conditions.

▼ **Figure 2.6** Example of streetscape lighting (courtesy of CU Phosco and TfL)

In street lighting, the recommended amount of light is quantified for different types of road (see BS 5489-1). Horizontal illuminance is used for subsidiary roads and paths, and for motorised traffic on conflict areas such as shopping streets, intersections, or roundabouts. For main roads (defined as motorised traffic on routes of medium to fast driving speeds) the recommendations are given in terms of road surface luminance, which is a measure of how much light is reflected off the road towards the driver. Such

© The Institution of Engineering and Technology

luminance is affected by the reflectance properties of the road as well as by the quantity of light falling on that surface.

The amount of light required depends on the class of the road. PD CEN/TR 13201-1 and BS 5489-1 give guidance on selecting the lighting class for road lighting, while BS EN 13201-2 provides performance requirements for each lighting class. For main roads, the class depends on how busy the road is (vehicle flow rates expressed as Annual Daily Traffic or ADT), traffic speeds and the frequency of junctions or intersections. For subsidiary roads and paths, the class depends on how busy the road or path is, and the ambient brightness of the location (for example whether it is rural or urban). Tables 2.1 and 2.2 give some examples of how the recommended levels can vary, although in all cases a full assessment using the methods in the relevant British Standards should be made when setting recommended levels.

▼ **Table 2.1** Typical performance requirements for motorised traffic on routes (v > 40mph).

Type of road	Road lighting class	Minimum average road surface luminance (cd/m²)
ADT > 40,000 vehicles, frequent junctions < 3km apart	M2	1.5
ADT between 7,000 and 40,000 vehicles, junctions > 3km apart	M3	1
ADT < 7,000	M4 high junction density	0.75
	M4 low junction density	0.5

▼ **Table 2.2** Typical performance requirements for conflict areas and subsidiary roads.

Type of road	Road lighting class	Minimum average illuminance (lux)
Traffic interchange on busy, fast main road	C2	20
Pedestrian crossing on busy route in town	C3	15
Subsidiary road in town, normal traffic flow	P3	7.5
Quiet residential road in country village	P5	3

NOTE: Road classes specified in BS EN 13201 tables are not yet fully updated in line with BS 5489 tables. See ILP PLG03 *Lighting for Subsidiary Roads* for further information on this topic.

Safety and security issues are paramount for pedestrian walkways and, particularly, for steps. Illuminance levels on steps should be slightly higher than the surrounding area but glare should be avoided.

There are additional recommendations in the form of minimum semi-cylindrical illuminances for pedestrian areas covered by CCTV for prosecution purposes to reduce crime risk and insecurity. The semi-cylindrical illuminance provides a measure of how much light falls on a pedestrian's face.

2.4.4 Area lighting for security

In order to ensure an appropriate level of security, large outdoor open areas, such as storage yards, are typically lit by means of floodlights mounted on construction elements or lighting columns. Additionally, lighting the perimeter of the outdoor area can also enable easier detection of intruders. Detailed guidance on area lighting is given in the CIBSE/SLL *Lighting Handbook*. The majority of light sources are suitable for these areas and in general the main driver is good light control to prevent obtrusive light as well as the inclusion of adaptive lighting and time-based control to reduce energy consumption. LEDs can be easily controlled and are not sensitive to frequent on-off switching. They can therefore be integrated into control schemes that incorporate presence detection and avoid the need for expensive HID instant restrike igniters.

2.4.5 Sports floodlighting

Sports floodlighting is a specialist area of lighting with complex criteria that are outside the scope of this Guide. However, its presence has an impact on other forms of exterior lighting and needs to be taken into consideration. The level of illuminance, uniformity, luminance contrast, colour temperature, colour rendering and glare control need to be considered carefully within the entire three-dimensional volume of the sports space above the field of play, while avoiding obtrusive light and light pollution.

The individual sport governing bodies often specify the required lighting design criteria depending on the whether the clubs are playing at local, regional or national levels. Consideration of television requirements often takes precedence over other requirements.

Further detailed guidance on sports lighting is given in the SLL Lighting Guide 4: *Sports Lighting* and requirements noted in BS EN 12193.

2.4.6 External architectural and amenity lighting

This category covers the exterior lighting of buildings and monuments, and lighting in parks and other public areas. Such lighting can have a major beneficial impact on the urban landscape and can enhance the architectural features of buildings as well as providing an impetus to extending commerce in the area (the '24 hour culture').

Traditionally, much architectural and amenity lighting has consisted of floodlighting objects and buildings by aiming beams of light at them. Floodlighting should achieve good modelling of architectural features but flat, blanket coverage of light without areas of light and shade is unflattering to most buildings. A typical strategy is to light the building façade at an angle so that there is a contrast between the architectural features and the background wall. The beam angle (i.e. the sideways spread) of the floodlight and the aiming angle should be chosen to give good illumination with minimal light spill to the sides or top. The level of illumination should distinguish the building or object from its neighbours without creating reflected glare. The colour of the lighting should match the materials being lit; higher colour temperatures where a cool, modern look is required, and lower colour temperatures to give warm floodlighting to stone or brick buildings.

Because of their small size, LEDs offer an alternative lighting strategy in which the light sources are integrated into the object to be lit. This can be more energy efficient, and enable intricate and striking lighting effects. It can be a particularly effective approach for open structures such as bridges.

LEDs offer a range of colours that can be used to give both static and changing colour effects on the object to be lit. This can be very eye-catching and can enable an environment creating a sense of place or providing an event space. However, this should be used sparingly and with due consideration to lighting requirements as a large expanse of coloured light may be unhelpful to people who are moving around or carrying out visual tasks in the area.

2.4.7 Illuminated signs

Illuminated signs are typically used outdoors as traffic signs or for advertising, and require sufficient brightness to stand out from their surroundings. For traffic signs, the illumination requirements should meet those set out in BS EN 12899 and the Traffic Signs Regulations and General Directions (TSRGD) legislation and are based upon the background ambient lighting levels.

For all types of signs, the light sources should be located so as to avoid uncomfortable reflected glare or shadows. Shielding might also be required for lamps that can be viewed by traffic facing the direction of the sign.

▼ **Figure 2.7** Example of illuminated signage (courtesy of Indo)

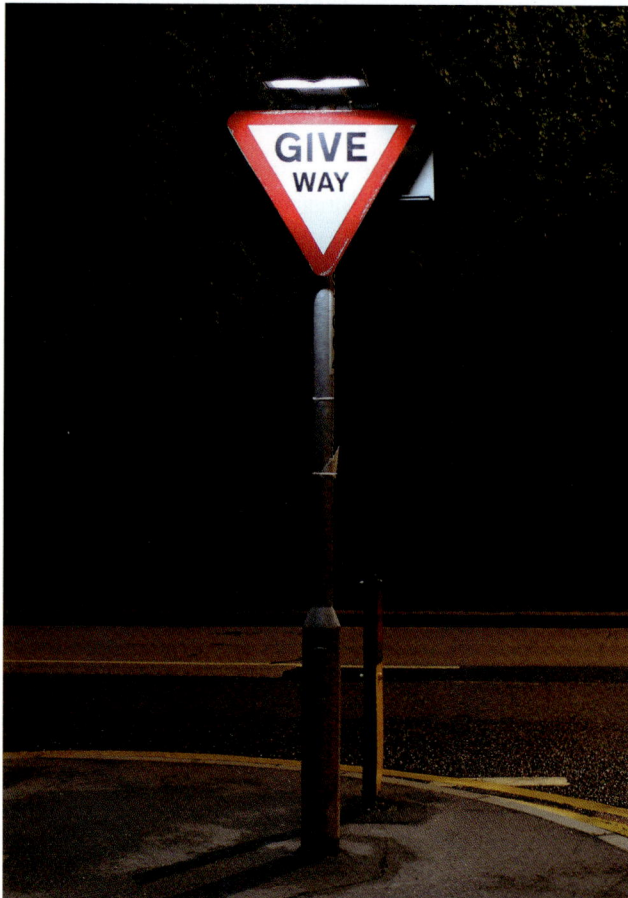

For advertising signage, careful aiming is needed to ensure uniformity and to avoid spill light. External illuminated signs need to appropriately protect against ingress of dust and moisture (see ILP PLG05 *The Brightness of Illuminated Advertisements*).

Technology change-over, retrofit and replacement considerations

3.1   Introduction

This Section reviews the practical considerations for energy-efficient lighting system change-over, including retrofitting or replacing existing stock with LED lighting technologies where applicable. Further coverage on key issues is included in Section 6 for performance requirements, Section 8 for selection criteria and Annex A for the Good Practice Specification Template (www.theiet.org/exterior-lighting) where specifying energy efficient exterior lighting systems.

3.2   Current and emerging technology

3.2.1   Technology development

Technology development for the exterior lighting market has seen improvements in light output and lifetime across a range of technologies, as outlined in Table 3.1 (with further detail on the history of technology development provided in Annex B).

▼ **Table 3.1** Exterior lighting technologies and typically quoted performance figures

Lighting technology	Typical light source luminous efficacy	Typical lifetime
Gas mantle	1 to 2 lm/W	-
Incandescent lamp	10-20 lm/W	1,000-2,000 hours
Fluorescent lamp	45 to 105 lm/W	6,000 to 15,000 hours
Low pressure sodium lamp	80 to 180 lm/W	10,000-18,000 hours
High pressure sodium lamp	45 to 130 lm/W	12,000-24,000 hours
Induction lamp	65 to 87 lm/W	80,000 to 100,000 hours
Metal halide lamp	60 to 130 lm/W	8,000 to 12,000 hours
LED retrofit lamp	40-80 lm/W	Up to 25,000 hours
LED Luminaire	100 to 130 lm/W	Up to 100,000 hours

NOTE: The lamps above have their luminous efficacy provided at the light source level. There will be additional losses of 25 to 40 % when these sources are included in a luminaire. As an example, for a typical high pressure sodium luminaire with 70 % LOR, the maximum luminous efficacy would reduce to (130 x 70/100) = 91 lm/W.

As illustrated in Figure 3.1, which shows the development of luminous efficacy for a range of light sources, the performance (luminous efficacy) of emerging technologies such as LED lighting have increased rapidly to the point of becoming good enough to enable viable and economic luminaires for exterior lighting applications.

▼ Figure 3.1 Luminous efficacy improvement for key lighting technologies

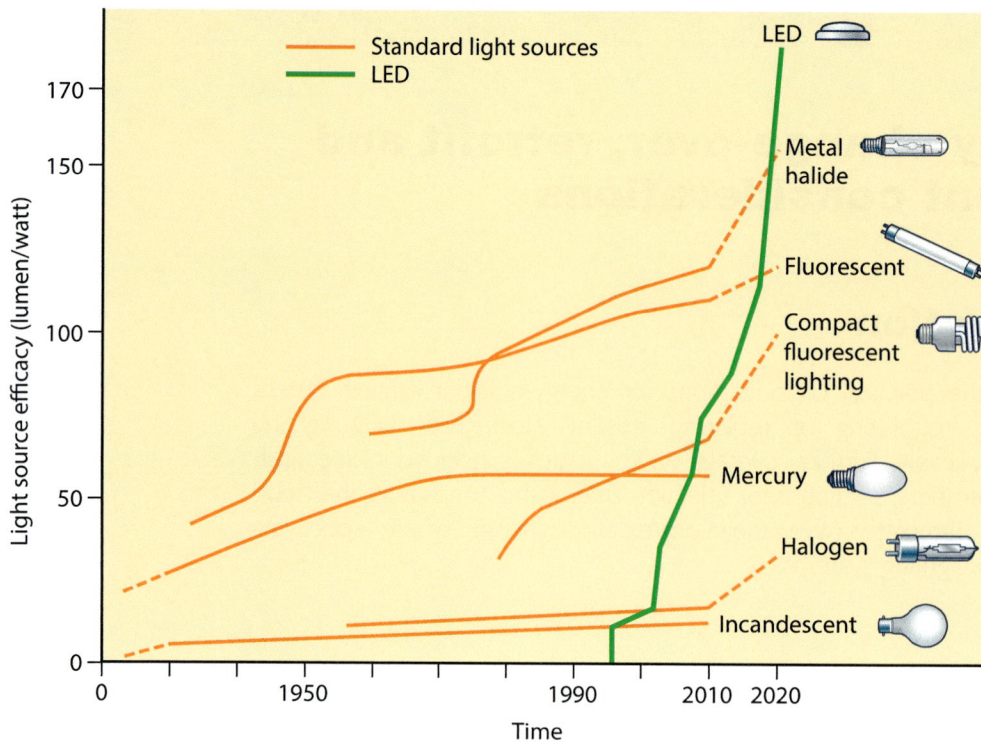

Further, perhaps more gradual, improvement in LED lighting technology is expected to continue over the coming years although future developments may focus on bringing the costs down rather than simply increasing performance, in order to support technology adoption.

3.2.2 Change management for exterior lighting systems

In making the case for change-over, asset managers and lighting designer need to compare suitable technologies, bearing in mind the differences between LEDs and conventional light sources.

Legislation is forcing an end to the production and/or import of some sources and this should be taken into account when looking at compliance issues. Product literature and articles in the technical and trade media will provide material for broad comparisons of characteristics such as optical/photometric performance, ranges of Correlated Colour Temperature and colour rendering, although the lighting industry has yet to settle on a metric that provides an accurate and complete comparison of the conventional and solid state sources' colour rendering.

The Good Practice Specification Template included in Annex A, and available to download online (www.theiet.org/exterior-lighting) provides a basis for specification of exterior lighting systems and the evaluation of supplier offers for suitable solutions. It is designed to be used to set out and assess criteria so that alternative offers and lighting solutions can be assessed on a like-for-like basis, i.e. so that the right type and amount of light is available for the required period of time and covers the required area in line with the appropriate standards.

Key considerations to note in comparison of conventional technologies with LED lighting systems include:

- that while luminous efficacy (lumens per Watt) provides a proxy for energy efficiency as a useful rule of thumb for an initial comparison, care should be taken not to confuse the efficacy of the luminaire with the efficacy of the source (lamp or LED).

- in terms of reliability, the most common failure mechanism for LEDs is different to conventional light sources, i.e. 'abrupt failure', since LED lighting systems commonly fail by 'parametric failure', a reduction in light output below the designed lighting level.
- LED lighting systems rely on complex electronics and so should be considered holistically in terms of design for particular applications.

Practical considerations in the comparison of technologies include the system performance, predicted life expectancy, reliability, routine maintenance and environmental benefits, are discussed below and in further detail in Section 6 on Performance requirements.

3.2.3 Adoption of LED lighting systems

At the time of writing, LED technology adoption for exterior lighting applications may best be described as being at the start of early majority adoption. LED lighting has become as increasingly attractive option for exterior illumination thanks to improvements in light output and luminous efficacy, advances in the design of drivers and controls, and a gradual reduction in the cost of luminaires.

▼ **Figure 3.2** Technology Adoption Lifecycle

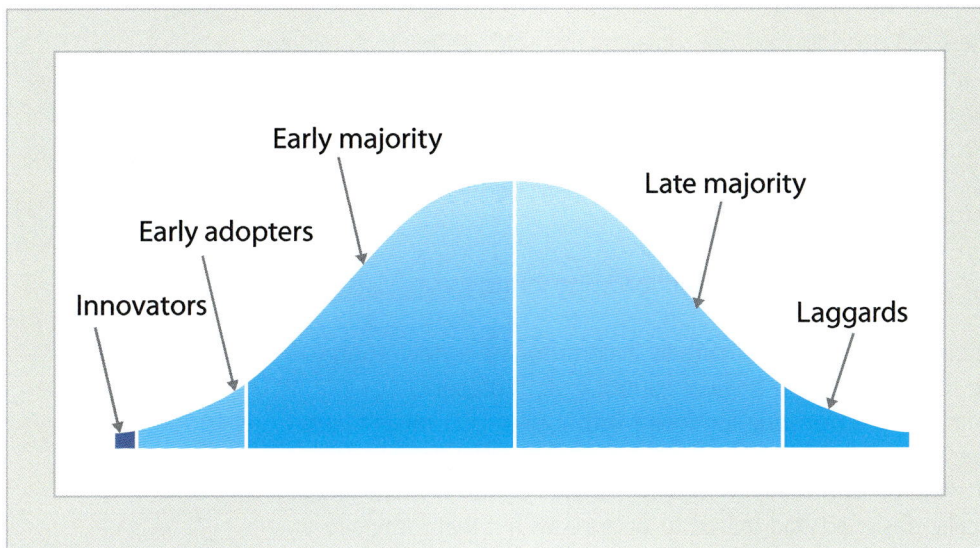

The environmental and operational benefits of moving to LEDs include:

- a reduction in raw materials, energy and waste materials compared to the production of conventional lamps
- a reduction of packaging materials and landfill
- less time spent by operatives on or near live carriageways carrying out maintenance
- adherence to the WEEE and RoHS Directives.

While there are many benefits to adoption of LED lighting systems if the specification is written properly and due diligence carried out, there remain:

- large variations in the quality of equipment and components in the marketplace.
- inconsistent labelling or product information which can make it difficult to compare systems performance.
- potential for sharp optical cut-off from some LED systems that may be considered too harsh for some applications.
- issues with weight and wind area, as LED luminaires are generally heavier than conventional lamp based luminaires.

- thermal management issues on some LED systems that can reduce light output, lifetime or both.

Effective adoption of LED lighting technology requires consideration not only of technology but also of the context of the area and the task requiring illumination. Prior to moving to technology adoption, it is important to first understand the rationale for a retrofit or replacement approach in order to meet the stakeholder's requirements and expectation. This should consider:

- the most effective way of saving money and energy with a lighting installation.
- whether lighting should be provided at all, and if lit, to what lighting class or designed level.
- the performance of previously installed products, including light distribution from the luminaire and colour characteristics.

3.3 Retrofit and replacement of exterior lighting systems

3.3.1 Retrofit and replacement options

Many lighting asset owners will need to consider replacing their old lighting technologies to benefit from energy and cost savings or because the existing system has come to the end of its operational life. Retrofit lighting systems, as distinct from replacement luminaires, may reduce energy consumption and make tangible savings, but it is important to have an understanding of retrofit options in order to ensure that the most effective solutions are provided.

Retrofit and replacement options generally take the following forms:

1   Retrofit lamp: retrofitting an existing luminaire to accommodate a new energy-efficient light source but retaining the original control gear, if present.
2   LED retrofit lamp: retrofitting an existing luminaire with a new LED retrofit lamp with an integral driver.
3   Light engine retrofit system: retrofitting an existing luminaire with a more efficient light engine, specifically designed and tested to operate within that luminaire and with compliant thermal management, EMC and electrical safety requirements to BS EN 60598 for luminaires and designed with optical performance and distributions to meet the relevant BS EN 13201 lighting class. The original lamp, gear, lamp holder and optical control are disconnected and discarded.
4   Replacing the entire luminaire with a new, specifically designed energy-efficient luminaire.

Any lighting solution, retrofit or otherwise, should be safe to operate. It is therefore essential to confirm that the proposed retrofit products operate safely within the specific luminaire in use and to show by testing that the resulting retrofit luminaire provides the equivalent or improved performance to the existing system, in particular, with regards to the distribution of light and its colour characteristics.

Test data that indicates the effectiveness and compatibility of the proposed retrofit solution in the final luminaire body – including ingress protection, electrical safety, thermal management, and optical control and EMC criteria – should be sourced from the relevant manufacturer. It is also recommended that trial installations should be performed and evaluated prior to further deployment.

3.3.2 Lamp retrofit solutions

Some LED retrofit solutions aim to provide a direct replacement for the original lamp source and are either designed to work with the existing control gear or require rewiring and/or the replacement of control gear. There is a large variation in the quality and effectiveness of this type of product, and retrofitting LED lamps into existing discharge lanterns is relatively rare.

The light distribution of direct replacement LED lamps is not usually the same as that of the original lamp. Following the retrofit of a luminaire, the lighting distribution is likely to be different and it must be reviewed to ensure it is acceptable. Where the LED retrofit lamp could be visible within a luminaire, the overall appearance of the retrofit solution should be considered in terms of aesthetics as well as colour appearance and any risk of glare to those nearby.

3.3.3 Light engine (lamp and control gear) retrofit solutions

LED engine retrofit kits can take various forms. Most follow a similar principle in that the luminaire body is retained, but the control gear, lamp holder and lamp are removed and replaced by an LED engine, which contains the optical, mechanical, electrical components, and all the associated electrical control gear to operate as a complete unit. LED engine retrofits that use projection technology can be more efficient than the reflection technology used with traditional lamps and reflectors (i.e. due to lower light loss within the luminaire). The lifetime and output performance of an LED engine retrofit solution is influenced by a number of factors including how effectively heat is dissipated away from the LED junction.

▼ **Figure 3.3** Integrated LED retrofit module solution, designed by the product manufacturer for retrofit and replacement of conventional discharge lamp solution (courtesy of DW Windsor)

3.3.4 Luminaire replacement

It is often effective to use a proprietary LED luminaire which is specifically designed for an LED source than to use a like-for-like retrofit product within an existing luminaire for exterior lighting applications.

If the lighting column is suitable to be retained (subject to checks on the limiting weight and wind area of the luminaire), the entire luminaire can be replaced with a new proprietary LED solution. The benefit of this approach is that optical and thermal control can be optimised to ensure a long life and high efficacy, subject to whole-life costings.

Where luminaires are being retrofitted to existing columns, it is essential that structural calculations are carried out to ensure that the weight and wind area of the luminaire can be accommodated by the existing structure. Should this not be the case, then the columns must be replaced.

3.3.5 Feasibility checks

It is important to understand the rationale for retrofit in each specific case. This not only includes an assessment of the most effective way of saving money and energy but also whether lighting should be provided at all and if lit, to what lighting class. Switching lights off when they are not required remains the most effective method of energy and cost saving.

To determine whether a change to more efficient lighting is appropriate (and, if so, whether retrofitting or replacement is the appropriate solution) it is important to understand what technology is being replaced, and why.

Before undertaking any intensive retrofit or replacement programme, it is advisable to visit the site to review the requirements and establish the most appropriate retrofit solution.

The considerations in Table 3.2 provide a sample checklist for making informed decisions when undertaking an energy efficient lighting system change-over.

▼ Table 3.2 Considerations for energy efficient lighting system change-over

Questions	Considerations
Why is new lighting required?	To save money? To save energy? To reduce maintenance and associated costs? To replace faulty equipment?
What site and operational criteria apply?	Photometric performance requirements Installation type? Hours of operations? Area covered/height of mounting? Energy costs?
What lighting systems are going to be replaced?	Condition of the lighting fixtures currently installed? What tasks are being performed in the space? Physical setting of lighting systems to be replaced? Any restrictions on luminaire mounting (for example height, weight, wind area)? Are the luminaires subject to unusual temperature ranges?
What control systems are going to be replaced?	Existing lighting control system? Adaptive lighting functionality?

A simple financial feasibility check for a retrofit solution is outlined in Figure 3.4, where a three-part calculation can be used to determine annual energy cost savings from upgrading one lamp or system type throughout a facility/site. The annual cost savings figure may be compared with the cost of the upgrade to determine simple payback and rate of return (ROR).

▼ **Figure 3.4** Example LED retrofit cost savings calculation

1. Calculate the total power (kilowatts, kW) potentially saved by replacing lamps

Original lamp wattage		Replacement lamp wattage		Energy saved per lamp		No of lamps to replace		Total watts saved	
[ ] W	–	[ ] W	=	[ ] W	×	[ ] lamps	=	[ ] W

Total watts saved			Total kilowatts (kW) saved	
[ ] W	*(divided by 1 000)*	=	[ ] kW

2. Calculate the total power (kilowatt hours, kWh) that could be saved annually

Total kilowatts (kW) saved		Hours of use per day		Days of use per week		Weeks of use per year		Total kWh saved per year	
[ ] kW	×	[ ] h	×	[ ] d	×	[ ] wks	=	[ ] kWh/yr

3. Calculate the total energy (kilowatt hours) that could be saved annually

Total kWh saved per year		Energy cost per kWh		Total energy cost savings per year
[ ] kWh/yr	×	£ [ ]	=	£ [ ]

$$\text{Payback} = \frac{\text{Initial cost of lighting upgrade}}{\text{Total energy cost savings per year}} = [\quad] \text{ years}$$

$$\text{Rate of return} = \frac{100}{\text{Payback}} = [\quad] \%$$

NOTE: For fluorescent or HID systems, substitute 'lamp' with 'system' or 'fixture' so that circuit watts are included.

Further annual costs savings can be made by participants of the CRC Energy Efficiency Scheme, based on calculating carbon allowance by multiplying annual kWh energy usage by the relevant power supply CO_2 conversion factor and the applicable CRC allowance. Further guidance is available from the Environment Agency. Furthermore, moving beyond annual cost savings to consider lifecycle costs such as reduced maintenance requirements can improve overall payback.

Wider feasibility checks may be required for lighting retrofit/replacement programmes in order to confirm total energy savings and/or whole life savings, based on, for example, load profiles, operating regimes, lighting power density indicator (W/lx/m²) and/or annual energy consumption indicator (Wh/m²) assessment criteria. Whole life savings is considered further in Section 4.

Financial management

4.1 Introduction

When financing or investing in energy reduction projects it is important to consider how to identify, develop and successfully bid for funds. This Section looks at this process for energy efficient exterior lighting projects, while potential options for public/private sector funding are reviewed in Annex C.

NOTE: The Street Lighting Toolkit (see Section 1.5) provides the means to model product ranges and operating regime codes for a local authority asset base.

4.2 Value engineering

Value engineering is a process of developing an optimal solution to operational needs by reducing inefficiencies in the design, construction, operation and maintenance.

Two priorities of value engineering are:

- understanding the current position; and,
- appraising options to identify the most appropriate solution(s) for the scheme development.

To be truly sustainable, a lighting installation should remain in place until a more efficient replacement solution is economically and environmentally viable. This point is reached when:

- energy and carbon savings make it cost effective;
- improvements in operational life/reliability reduce reactive and planned maintenance activities;
- previous 'invest to save' funding has delivered; and,
- payback and benefit have both been achieved.

4.3 Understanding the current position

4.3.1 Determining the base line

Any organisation looking for funding to promote energy and operational savings will need to demonstrate an understanding of their current assets, based on an accurate and well-maintained inventory data. A lighting asset owner should be able to assess the current performance and energy consumption of its lighting and use this information to develop energy reduction strategies and improve their operation. The asset database/inventory system will provide the core data needed to measure energy savings and other efficiencies. The inventory should be used to manage maintenance and provides a record of the history and condition of assets.

The ILP document 'Managing Unmetered Energy Street Lighting Inventories' provides detailed guidance on setting up and checking an inventory for unmetered street lighting assets and can be downloaded from https://www.theilp.org.uk/home/

4.3.2 Calculation of baseline energy use

It is important to quantify current energy use in order to forecast energy/carbon savings and create the associated cost-saving model. For local authorities, let us assume that an asset owner has appointed a meter administrator (MA) and is trading its energy on a dynamic half-hourly basis. The MA is responsible for providing half-hourly consumption data in settlement periods based on inventory data and burning hours: in other words, consumption in kWh, for each half-hour of every day.

The annual energy consumption is calculated using the following formula:

Estimated annual consumption (kWh) = (Circuit Watts × Annual burning hours × Number of units) ÷ 1,000

Where lighting installations are subject to metered connections then the baseline energy consumption should be referenced to actual past meter readings where this is possible, or calculated costs base on circuit watts.

4.3.3 Operational costs

It is likely that any energy efficiency programmes that are instigated by a lighting asset owner will also bring operational improvements. For example, automatic monitoring will remove the need for scouting to check whether lamps are operational or not; improved operational life will result in changes in lamp clean-and-change regimes, and increased reliability will reduce the level of reactive maintenance operations. These savings may even assist the asset owner fund new lighting infrastructure.

4.3.4 Calculation of forecast energy and carbon

Energy savings are calculated using the fundamental principle of comparing energy use before (baseline) and a forecast for after the new technologies (such as LEDs) are installed, by identifying the relative saving. The carbon savings can then be calculated to determine the tonnes of CO_2 saved against the Carbon Reduction Commitment Energy Efficiency (CRCEE) scheme value.

4.4 Investment options appraisal

Reductions in energy consumption reduction are not achieved with new technology alone but also through the design, operational and maintenance aspects of a project. For example, a CMS does not bring savings in itself, but it facilitates energy and operational savings through control and monitoring.

When appraising technical options it is important to consider the investment level compared against the life of the scheme, the lighting master plan, guidance specific to the project location, and/or technology development. For example:

- a low-cost, shorter life scheme may be more beneficial than a higher-cost, longer life scheme that is less flexible in terms of adapting in line with new technology.
- a scheme may or may not need to include column replacement.
- a scheme may rule out or include full CMS, trimming or adaptive lighting as options.

Each individual project appraisal can include any or all of the different options including or excluding as necessary column replacement, intelligent controls, and varying product life options.

© The Institution of Engineering and Technology

Investment option	Sample considerations
Do Nothing	- Assets remain in place and continue to deteriorate until they fail to provide any benefit of lighting.
Do Minimum	- Minimum investment required to get existing equipment back into working order to provide some limited benefits from lighting. - State if scheme is to be fully compliant with standards or just 'in light'.
Low Investment	- Partial/full retrofit scheme for luminaires and or intelligent controls on a 'one for one' replacement with no planned column replacement. - Controls include simple trimming and part-night switching through intelligent photocells with no dimming provided. - Further capital replacement will be required to replace end of life column assets. - Luminaire replacement required no sooner than 5 years.
Medium Investment	- 'One for one' retrofit luminaires and where necessary new column replacement bringing assets up to a minimum 10 year remaining life. - Controls include trimming and part-night dimming. - Luminaires provide full dimming controlled via intelligent drivers/photocells, used throughout.
High Investment	- Full luminaire and column replacement scheme using relocated column positions at optimised spacings. - Full CMS control systems. - Scheme life 20 years minimum.

Applications for funding need to show that all aspects have been duly considered and that options have been tested, at an early stage and using a whole-life costing approach (see below).

Sources of information include the Energy Saving Trust (www.energysavingtrust.org.uk).

4.5 Realising whole-life savings

4.5.1 Technologies

Where possible, asset owners should liaise with others undertaking similar programmes, in order to understand the technologies and operational strategies they propose to use. Sharing lessons learned is valuable in itself but also helps to demonstrate thoroughness in the process of vetting and assessing solutions. On-site trials should follow a desk-based review of technologies, for similar reasons. The energy performance of any proposed lighting equipment should be included in the inventory.

4.5.2 Design

Applying good practice in lighting design in line with standards such as BS EN 13201 and national guidance forms a critical element in achieving energy-efficient lighting solutions. Within good design, it may be possible to reduce costs by increasing the distance between the streetlights and/or reducing lamp power, but this has to be balanced against the capital costs and requirements for new columns and electrical connections. A one-for-one upgrade could provide a more cost-effective solution so long as the lighting designer ensures that the column is suitable, structurally, for the new bracket and/or luminaire.

The asset owner should support the overall forecast energy savings for the project with specific energy savings based on representative proposed lighting calculations. For local authorities, this would include using the standard energy format defined by Elexon and identifying any adaptive lighting profiles. This information should show the assumptions and the underlying calculations used. Although it is not a requirement of the Green Investment Bank's Green Loan financing, it may be appropriate to note operational revenue savings in both planned and reactive maintenance, and savings under the CRCEE scheme.

4.5.3 Whole-life costing

In the past, projects tended to be assessed on the construction costs (capital works) without taking into account the operational, maintenance and end-of-life costs associated with the project. But what might appear to be low-cost installations might produce higher long term operational and maintenance costs. By contrast, whole-life costing is used to assess project options based upon capital, operational, maintenance and end-of-life costs with the aim of providing the most cost effective installation throughout its life.

4.5.4 Payback

Payback is the length of time required to recover the cost of an investment. The payback period of a given investment or project is an important determinant of whether to undertake the position or project, as, for instance, local authorities generally do not favour investments with longer payback periods.

Payback Period = Cost of project ÷ Annual cash inflows

It should be noted that payback is the point at which the cost of investment breaks even with savings. All clients should remember any existing debts on recent invest to save projects if they intend to undertake additional investment.

NOTE: Payback is one option for assessment but clients should consider what assessment options are relevant to them.

© The Institution of Engineering and Technology

Asset management

5.1 Introduction

The Institute of Asset Management describes asset management as 'the balancing of costs, opportunities and risks against the desired performance of assets, to achieve organizational objectives'. This Section looks at fact-based asset management in terms of the significant investment local authorities and other asset owners make in exterior lighting over the whole life of lighting systems from design to replacement.

5.2 Role of asset management

5.2.1 General

Asset management is essentially about understanding what you have so that you can manage your service and infrastructure to ensure that it performs as required, is in a safe condition and can be developed for the future. It may be defined as 'knowledge-based planning, focused on outcomes'.

5.2.2 Knowledge

Accurate and up-to-date knowledge of assets and the associated attributes (for example, light source electricity costs, Elexon UMSUG codes for unmetered supplies, materials, dimensions, condition, installation life and costs) should be held in a user-friendly database that can be readily accessed for reports. It is important to maintain the quality of this data through formal information management activities, including data collection programmes, routine maintenance operations and audits.

5.2.3 Planning

The whole life of assets should be considered when planning works, with a view to finding the most cost-effective solution that meets service level requirements. This approach produces long-term financial and work forecasts that, once agreed, are documented in the asset management plan. An example of this approach is discussed in Textbox 5.3 which outlines considerations for a Lighting Value Management Model (LVMM). These long term planning techniques are used to bid for and allocate budgets objectively and optimally, and also allows for the impact of different levels of funding on the service to be assessed.

5.2.4 Outcomes

Work planning focuses on delivering outcomes, such as improved lighting performance, to the customer, which might be the public.

5.2.5 Asset database/inventory system

The bedrock of this process is the asset database/inventory system. The inventory enables asset owners to manage the Unmetered Energy (in the case of local authorities), organise and carry out maintenance and record the history and condition of assets, and thus informs its ability to upgrade. The inventory not only provides the information required to manage the lighting service effectively, but also provides the core data needed to develop energy and carbon reduction strategies.

Textbox 5.1 Experience from the Highways Maintenance Efficiency Programme

UK Roads Liaison Group has this to say about asset management, and while this relates to highway lighting it can equally be applied to private sector lighting installations:

Asset management has been widely accepted by central and local government as a means to deliver a more efficient and effective approach to management of highway infrastructure assets through longer term planning, ensuring that standards are defined and achievable for available budgets. It also supports making the case for funding and better communication with stakeholders, facilitating a greater understanding of the contribution highway infrastructure assets make to economic growth and the needs of local communities. In England, the Highways Maintenance Efficiency Programme (HMEP) has recognised that better advice and information is required if asset owners are to benefit consistently from the potential that asset management offers.

▼ **Table 5.1** Components of good asset management

Context	Asset Management Policy and Strategy	Risk Management	Performance Management Framework	Leadership and Commitment
Highway Infrastructure Asset Management Framework	Asset Management Systems	Lifecycle Plans	Performance Monitoring	Communications
Making the Case for Asset Management	Asset Data Management	Works Programming	Benchmarking	Competencies and Training

NOTE: components of good asset management should not be considered in isolation.

NOTE: These components of good asset management are covered in detail within the HMEP UKRLG *Highway Infrastructure Asset Management Guidance* and as such will not be discussed here.

5.3 Benefits of asset management

In general, the adoption of asset management enables the asset owner to provide the same or better level of service at a reduced cost, or a better level of service for the same or marginally increased cost.

NOTE: It will be important for local authorities to demonstrate no loss of service provision in order to align with Gershon Efficiency Savings in the UK public sector.

Other, more specific benefits of adopting an asset management approach include:

- improved capability to demonstrate and justify the funding required for service delivery;
- improved knowledge of whole-life needs enabling costs to be minimised;
- improved linkage between annual maintenance work and long-term strategic vision;
- improved visibility and communication of performance, both internally and externally;
- improved decision-making through better information and objective resource allocation;
- improved risk management through explicit identification and analysis of risks;
- improved financial control through better understanding of costs and expenditure; and
- improved value for money and management control by using transparent, auditable, robust and reliable information, processes and systems.

5.4 Scope of asset management

The broad scope of asset management functions can be described in the three levels of strategic, tactical and operational, as summarised below:

▼ **Table 5.2** Asset management levels and functions

Level	Functions
Strategic: where are we going and why?	- An organisation establishes its overall long-term direction, for example, policy, goals and objectives, vision, mission statement and targets.
Tactical: what is worth doing and when?	- The asset managers translate the strategic goals and objectives into specific plans and performance targets for individual asset types, for example, lighting. - A performance gap analysis and a formal planning process are applied to identify the required, most beneficial and cost-effective activities and when they should be carried out. **NOTE:** The development of the HMEP for public authorities is a tactical activity.
Operational: how to do the right things?	- The asset managers, engineers, technicians and operatives develop and implement detailed work plans and schedules that have a short-term outlook but take account of the work volumes and phasing, for example, arising from the HMEP. - Engineering processes include inspection, routine maintenance, scheme design, work scheduling and implementation and will be recorded for example, within a Lighting Maintenance Management Plan (LMMP).

Textbox 5.2 Lighting Maintenance Management Plan (LMMP)

The production of a Lighting Maintenance Management Plan (LMMP) is not a statutory requirement. However, the LMMP follows national good practice, shared by many local highway authorities in detailing policy, strategy and operation of street lighting maintenance, in a single comprehensive document. The plan will become an integral element of an asset management plan, which itself may become a statutory document required from all local highway authorities in England in the near future.

'Well-lit Highways' - *Code of Practice for Street Lighting Maintenance Management* is based on the principles of best value and continuous improvement and is an important component of a Transport Asset Management Plan. It is also equally applicable to non-highway lighting installations managed by the private and public sectors. Services should be based on the needs of users and the community rather than the convenience of service providers; hence the need for local flexibility.

The London Lighting Engineers Group (LoLEG) have produced a framework document that looks at the considerations to be made when developing an LMMP; see: http://www.loleg.co.uk/

5.5   Lifecycle planning

5.5.1 Purpose of lifecycle planning

Lifecycle planning aids in the identification of long term investments for lighting infrastructure assets and the development of an appropriate maintenance strategy. It also permits the lighting asset owner to:

- predict future performance of the infrastructure assets for different levels of investment and different maintenance strategies;
- determine the level of investment necessary to achieve the required performance;
- determine the performance that will be achieved for available funding and/or future investment;
- support decision making and the case for investing in maintenance activities and demonstrate the impact of different funding scenarios; and,
- minimise costs over the lifecycle while maintaining the required performance.

5.5.2 Lighting maintenance policies and practices

The purpose of maintenance planning and management is to develop and implement cost-effective and sustainable maintenance plans that ensure that the lighting infrastructure remains in safe operation and performance. Maintenance policies and practices should be sufficiently flexible to respond to local circumstances but also consistent, particularly where the lighting is part of a bigger network.

NOTE: While the recommendations in *Well-Lit Highways* are not mandatory for local authorities, they include the key principle that a local authority – or other asset owner – should involve users in the design and delivery of its services, using reasonable discretion.

5.5.3 Asset data for lifecycle planning

Asset data for lifecycle planning should be available from an asset management system, asset register or maintenance management system. Typically, the following is required to develop lifecycle plans:

- inventory (road lengths, widths, area being lit, structural components and dimensions, lighting column types and sizes as a minimum);
- performance (including asset condition);
- routine maintenance (including reactive and cyclical maintenance activities); and,
- treatment options (including their historic performance and cost).

Well Lit Highways (Appendix A) details the core list of data sets that should be recorded against each street lighting asset and is just as applicable to any other lighting installation.

5.5.4 The lifecycle plan

Lifecycle planning comprises the approach to the maintenance of an asset from construction to disposal. It is the prediction of the future performance of an asset, or a group of assets, based on investment scenarios and maintenance strategies. The lifecycle plan is the documented output from this process.

Lifecycle plans may be used to demonstrate how funding and/or performance requirements are achieved through appropriate maintenance strategies with the objective of minimising expenditure, while providing the required performance over a specified period of time.

Lifecycle planning can be applied to all exterior lighting infrastructure assets. Its application may be more beneficial to those assets that have the greatest value, require considerable funding, are high risk and/or are seen as critical assets.

5.6 Future technology development

Public lighting is a long-life asset with columns having a design life of 25 plus years, luminaires 20 plus and light sources four-plus years. It should be managed with this design life in mind; the time scales involved make it important to allow for innovations and new equipment to be accommodated at a later stage without causing problems.

Textbox 5.3 Lighting Value Management Model (LVMM)

The Lighting Value Management Model (LVMM) is a risk management tool and a means of prioritising the street lighting budget spend through a ranking system for every street within a local authority area. The prioritisation process balances safety, cost savings with other considerations.

The LVMM calculates and prioritises the highest risk columns and lighting installations at any given time, taking into account factors such as structural condition, crime data, lighting performance and maintenance history. It is intended to maximise the impact on the annual budget and provide the best value returns for a lighting replacement prioritisation programme. The prioritisation process takes into account all the factors relating to the renewal of an existing lighting scheme, and by appropriate weighting of these factors, gives the installations that are most in need of replacement the highest listing of priority.

The production of an LVMM is not a statutory requirement. However a LVMM can be a key tool for the street lighting professional and would be an integral element of an asset management plan, which may soon become a statutory document for all highway authorities in England.

The London Lighting Engineers Group (LoLEG) has produced a framework document that looks at the considerations to be made when developing an LVMM; see: http://www.loleg.co.uk/

SECTION 6

Performance requirements

6.1 Introduction

This Section covers performance requirements in general terms and should be read in conjunction with the Good Practice Specification Template provided in Annex A, and available to download as an Excel spreadsheet (www.theiet.org/exterior-lighting). For further detail and for LED lighting systems in particular, reference should be made to the IET *Code of Practice for the Application of LED Lighting Systems*.

6.2 Lifetime, performance, failure

6.2.1 Lifetime and specification

Lighting performance requirements – typically the lighting parameters required at the end of the design life – should be established in accordance with the relevant standards and guidance. Assumptions should be documented to allow different lighting solutions to be compared and assessed.

The Good Practice Specification Template has been produced to enable designers and specifiers to concentrate on specifying output requirement (thereby facilitating competition and innovation), while enabling a meaningful comparison of proposed lighting solutions. This includes a number of items to ensure that the products and solutions offered are fit for purpose.

In order to ensure that the specification and procurement process is robust and appropriate, it is necessary to specify the relevant geometry (for example, for road lighting or area lighting) and the required lighting parameters including the lighting scheme design life. This will then ensure that different design solutions are appropriate.

6.2.2 Understanding lifetime and failure

In respect of the lighting scheme design life, the definition of 'lifetime' for lighting sources has two characteristics:

- a gradual reduction in light output with time, known as lumen depreciation.
- a known or calculated rate at which physical failures occur.

These are often shown by typical curves below.

▼ **Figure 6.1a** Typical lumen depreciation curve

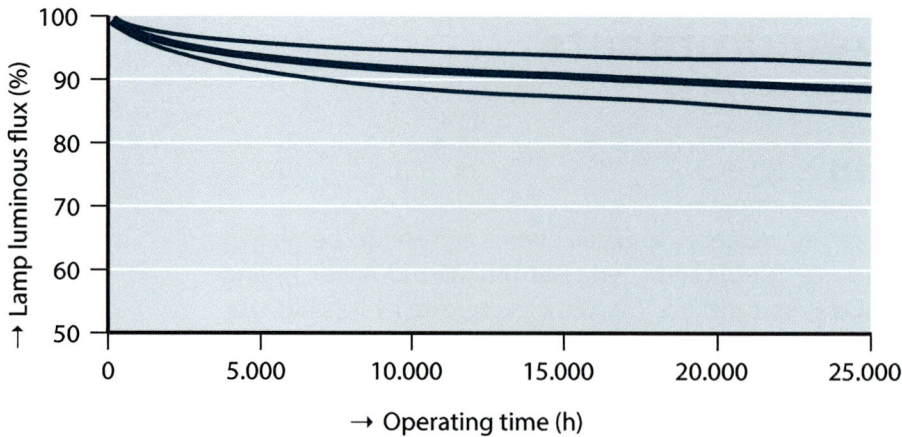

▼ **Figure 6.1b** Typical physical failure curve

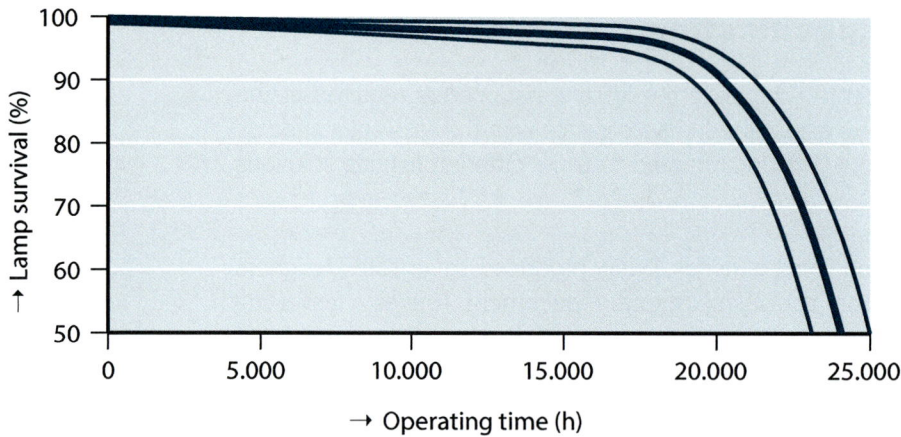

For conventional lighting sources, physical failure is used as the basis for the claimed lifetime, with lumen depreciation provided as additional information.

For LED lighting sources the lumen depreciation is used as the basis for the claimed lifetime, with physical failure provided as additional information. The Median Useful Life (Lx) parameter is the time for 50 % of a batch of LED products to fade to a given percentage of its original light output (under constant drive current conditions).

Usually, an L80 value is quoted meaning the time to drop to 80 % of the initial light output. L80 is deemed appropriate for many functional lighting applications as the light output value beyond this point may become unsuitable for purpose. Similarly, other Lx values are deemed appropriate for different applications, such as L50 (i.e. operating hours to 50 % of initial output) for decorative or accent lighting applications where the lighting intensity is less critical than the placement and distribution of light.

The Useful Life (LxBy) parameter may also be quoted where the percentage of the batch that fade below x % takes another value, for example, L80B10.

Alongside the Lx parameter is the Abrupt Failure Value (AFV). This is the percentage of LED light sources or luminaires of the same type that no longer give any light at the

Median Useful Life. For example, the Rated Median Life may be 50,000 hours for lumen depreciation to 80 % of the initial light output. At this time, the number of abrupt failures could be 6 %. This would be written as $L_{80} = 50,000$ AFV $= 6$ %.

For non-professional applications the two life parameters (Lx and AFV) are sometimes combined giving the Combined Failure Value (CFV).

The end of life of a lighting system is usually determined by its weakest component – where this cannot be easily and cost effectively replaced. In particular for LED lighting systems, failures in components other than the LED chip itself (for example, luminaire mechanical components or driver electronic components) may mean that, where practicable, such components need replacing during the design life of the installation. Where this is not possible or practical, such failures would create an end of life scenario.

In respect of 'lifetime', what is critical is the lighting system lumen output and relevant lighting design parameters at the end of the lighting design life. If these are suitable and the design criteria are met at this stage then, prior to that, the lighting design parameters will also be met or exceeded.

As noted in Section 3, where luminaires are being retrofitted onto existing columns, it is essential that structural calculations are carried out to ensure that the weight and wind area of the luminaire can be accommodated by the existing structure. Should this not be the case, then the columns must be replaced.

6.2.3 Design working life

The design working lives of luminaires will vary by manufacturer and product quality. For exterior HID lighting, luminaire life was historically 20-25 years. With luminaires now including more electronics, product life is more typically expressed in operating hours, for example, 50,000 hours to 100,000 hours. For 12 hour, 365 day operation the annual burning hours is approximately 4200 hours (the theoretical 4380 hours varies with location around the country) this gives product life expectancy of 12 to 24 years. Luminaires claiming life figures below or significantly above this 12 to 24 year range should be treated with caution and evidence requested to justify any claims.

6.2.4 Luminaire failure rates

The bathtub curve of failures, shown in Figure 6.2 describes the expected failure rates of a large population of electronic products. The first period is characterised by a decreasing failure rate and consists of failures caused by defects in materials or workmanship. These failures are usually covered by commercial warranties. A standard one-year warranty should provide sufficient time to uncover these initial defects and allow replacement or repair by the manufacturer.

The second period maintains a low and relatively constant failure rate and consists of random failures typically caused by 'stress exceeding strength.' These failures might relate to lightning strikes or electrical supply surges that cause damage to the electrical circuits in the driver or to the LEDs. Maintenance may be required to the equipment to ensure correct functionality for example, replacing perished gaskets to maintain ingress protection, but this should be economic to carry out for the asset owner provided that suitable spare parts are available from the manufacturer. Extended warranties are sometimes sold to cover the risks associated with unexpected maintenance requirements although these wouldn't normally cover the normal wear and tear failures due to, for example, lightning strikes and electrical surges.

The bathtub curve
Hypothetical failure rate versus time

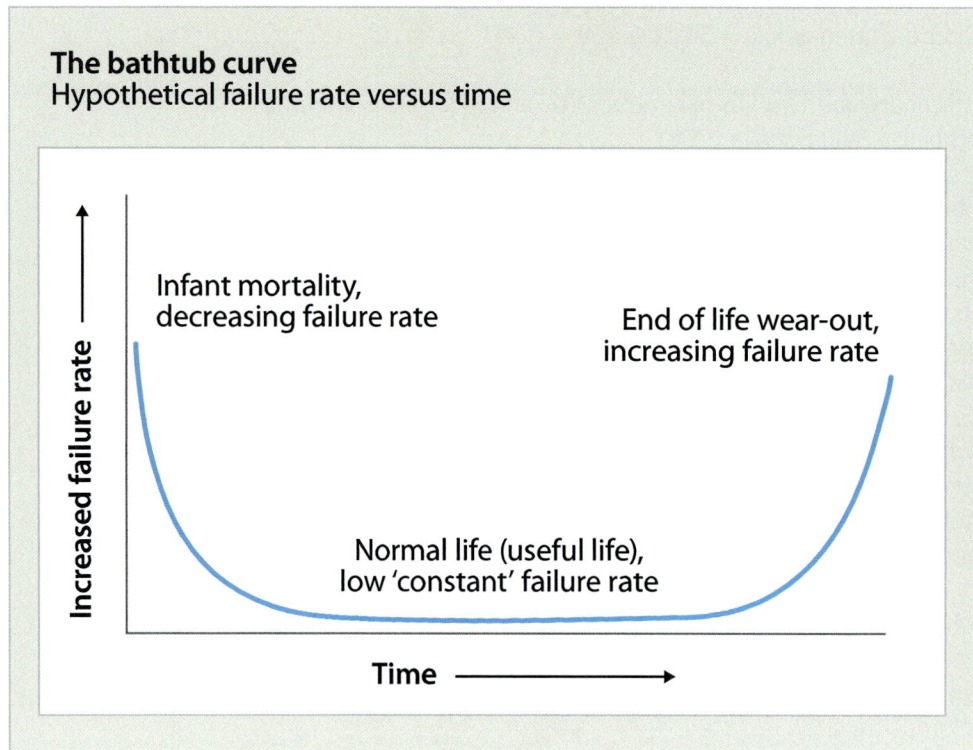

The third period exhibits an increasing failure rate and consists of failures caused by wear-out due to fatigue or depletion of materials, for example, the deterioration and failure of electrolytic capacitors in the electronics. These failures will be distributed either side of the actual end of life, typically defined as 50 % failure of the population. The actual end of life defined by 50 % failures of the luminaire population can only be determined statistically by monitoring the rate of failures over time and identifying the maximum failure rate, indicated by the point at which end of life failure rates start to decrease.

6.2.5 Actual working life

Actual working lifespan of the reliable luminaires that pass the 'infant mortality' period will be governed by a number of factors including the environmental conditions, the quality of the luminaire design and components used and the quality of work:

- The environmental conditions in which the luminaire is installed and operated affect the life of the luminaire for example, lightning strike risk levels, high ambient temperatures, high humidity, high air pollution levels, and high salinity all have the potential to adversely affect luminaires.
- The quality of the luminaire and its components, the thermal management provided within the luminaire, the protection of the luminaire enclosures against water ingress, the resistance to electrical surges, and the life of complex controls such CMS nodes, can all affect the luminaire life. The use of lower quality components of unknown origin may limit the life of luminaires significantly. The availability of spare parts will enable extension of the working life of the individual luminaires through maintenance and repairs.
- After installation, the competence and quality of workmanship of maintenance operations on the luminaire may have a significant effect on the life of the luminaire. Poor workmanship may encourage early failure of the luminaire through, for example, electrical failure caused by incorrect re-wiring, or replacement with incompatible components; or water ingress from compromised gaskets seals or failure to properly secure catches and fasteners.

6.3 System performance

6.3.1 Luminous efficacy

Luminous efficacy is a term normally used to describe the performance of a lighting system comparing the output to the input. The efficacy of a light source is given by the luminous flux it emits divided by the power consumed, and is expressed in lumens per Watt (lm/W).

It is worth noting that while a light source may have a very high efficacy under laboratory conditions, its efficacy can be much lower when the same light source is incorporated into a luminaire and operated with a driver. Also: at times the efficacy of a lighting system is listed as the efficacy of the light source only, but this can be misleading as it does not consider the losses presented by the other associated components within a luminaire.

6.3.2 Energy efficiency

For light sources to be considered energy efficient, the lighting solution metrics must also contain light output per unit of electricity consumed. For road lighting applications this is covered by two metrics which should be used together:

- Power Density Indicator (PDI or D_P) measured in $W/lx/m^2$, which is the value of the system power divided by the product of the surface area to be lit and the calculated maintained average illuminance on this area.
- Annual Energy Consumption Indicator (AECI or D_E) measured in Wh/m^2, which is the total electrical energy consumed by a lighting installation day and night throughout a year in proportion to the total area to be illuminated by the lighting installation.

6.4 Physical performance characteristics

6.4.1 Thermal management

Light sources and their control gear operate best at optimum temperatures, although as a general rule, electronic components operate best at lower temperatures. Thermal management is therefore critically important for LED light systems.

The luminaire must be designed to ensure appropriate thermal management, taking into account thermal paths via conduction, convection and radiation. Thermal management can be evidenced by the provision of thermal imaging and associated data.

Lighting systems should be provided with specification data that clearly defines their performance at given thermal ranges and ambient temperatures.

6.4.2 Physical protection

IP (or ingress protection) ratings are provided in BS EN 60529. They are used to categorise levels of sealing effectiveness of electrical enclosures against intrusion from foreign bodies (tools, dirt etc.) and moisture. Generally the higher the IP rating the better protected the enclosure is from external environmental influences and the less cleaning of the inside of the enclosure is required.

IK ratings show degrees of protection provided by enclosures for equipment against external mechanical impacts in accordance with BS EN 62262. The higher the IK rating, the greater the resistance to mechanical impact.

For most exterior lighting applications, IP ratings are more important than IK ratings, except where luminaires are mounted within arm's reach and may be susceptible to vandalism or areas considered to be hazardous zones.

Some of the most common IP ratings applicable to LED systems are indicated in Table 6.1.

▼ **Table 6.1** Common ingress protection ratings

Solid objects Protected against		Moisture Protected against	
IP0x	No special protection	IP0x	No special protection
IP1x	Objects larger than 50 mm, but not protected against deliberate contact with a body part	IP1x	Dripping water (vertical drops)
IP2x	Objects larger than 12.5 mm, such as fingers	IP2x	Vertically dripping water, when the item is tilted at angle of 15°
IP3x	Objects larger than 2.5 mm, such as tools/thick wires etc.	IP3x	Spraying water, when the item is tilted at angle of up to 60° to vertical
IP4x	Objects larger than 1 mm, most wires/screws etc.	IP4x	Splashproof in all directions
IP5x	Dust protected; not completely sealed but will not allow enough matter ingress to damage the item	IP5x	Water jets (6.5 mm nozzle) in all directions
IP6x	Dust tight	IP6x	Powerful water jets (12.5 mm nozzle) in all directions
		IP7x	Temporary immersion in water up to 1m
		IP8x	Continuous immersion in water deeper than 1 m (depth will be specified by manufacturer)

▼ **Table 6.2** Common mechanical impact protection (IK) ratings

IK	Impact energy (joules)	Equivalent impact
00	Unprotected	No test
01	0.15 J	Drop of 200 g object from 7.5 cm height
02	0.2 J	Drop of 200 g object from 10 cm height
03	0.35 J	Drop of 200 g object from 17.5 cm height
04	0.65 J	Drop of 200 g object from 25 cm height
05	0.7 J	Drop of 200 g object from 35 cm height
06	1 J	Drop of 500 g object from 20 cm height
07	2 J	Drop of 500 g object from 40 cm height
08	5 J	Drop of 1.7 kg object from 29.5 cm height
09	10 J	Drop of 5 kg object from 20 cm height
10	20 J	Drop of 5 kg object from 40 cm height
10+	40 J	Drop of 10 kg object from 40 cm height

© The Institution of Engineering and Technology

It is important to check the IP ratings of all LED system components (including any remote drivers that may not be contained within the LED luminaire, but may need to be located nearby).

There may be added benefits to specifying products with a higher IP rating in certain applications, in order to reduce the impact on system performance due to ingress of dirt/dust (i.e., in locations where regular access for cleaning may be difficult).

6.5 Electrical characteristics

6.5.1 Rated input power and driver performance

The rated input power of a luminaire (i.e. lamp and control gear/driver) is measured in Watts and can be used to calculate energy consumption based on operational profile (hours). The power input of a lamp is regulated by the control gear/driver, typically via constant drive current, though this can vary depending on the driver condition or through the operation of compatible lighting controls that adapt light output in accordance with designed lighting levels.

The driver also consumes power and typically may have an efficiency of approximately 85 % which reduces the efficacy of the system. Parasitic loads (such as driver input power in standby mode even when the lamp is switched) should also be considered in calculating overall energy consumption.

The lifetime of a driver is limited to the shortest-lifetime component based on the conditions in which it is running, such as the ambient temperature, power transients (spikes and surges) on the supply, and the demands placed on the driver by the output load (whether it is fully loaded or not).

The two weakest points in a typical driver are:

- the input circuit, which has to be able to cope with mains-borne transients that may enter the driver circuit; and,
- electrolytic capacitors in the output circuit of the driver.

For retrofit lamp solutions in particular, it is important to check the maximum rated wattage, the minimum load, and the lifetime of the driver/control gear to ensure it is appropriate for the designed operation of the connected light source.

NOTE: For LED lighting systems, too little current and voltage results in little or no light, and too much current and voltage can damage the light source.

6.5.2 Power factor

Power factor is the ratio between the useful (true) power (kW) to the total (apparent) power (kVA) consumed by an item of a.c. electrical equipment or a complete electrical installation. It is a measure of how efficiently electrical power is converted into useful work output.

The ideal power factor is unity, or one. Anything less than one means that extra power is required to achieve the actual task at hand. All current flow causes losses both in the supply and distribution system. A load with a power factor of 1.0 results in the most efficient loading of the supply. A load with a power factor of, say, 0.8, results in much higher losses in the supply system and a higher bill for the asset owner.

For unmetered supplies (i.e. the majority of road lighting in the UK), power factor correction is required of not less than 0.85.

6.6 Health & Safety information

6.6.1 Provision of information

Luminaires and lighting systems in general have to meet health and safety legislative requirements in the UK. Those placing luminaires on the market (i.e. manufacturers or distributors) have an obligation to provide information on the installation and maintenance of their products and to provide any necessary health and safety information (such as CE marking compliance).

6.6.2 CE Marking

CE marking requirements are included within a number of EU Directives, including those for low voltage electrical equipment, electromagnetic compatibility (EMC) and Ecodesign requirements for energy-related products and, where applicable, construction products regulations. Compliance is indicated on lighting system products via an affixed CE mark, but this alone should never be relied upon to ensure compliance. Those placing the item on the market should be able to supply the underlying declarations of conformity/declarations of performance in accordance with the EU Directives and where applicable the Member State National Rules. For luminaires, compliance with BS EN 60598 *Luminaires. Particular requirements* may be accepted as meeting the relevant CE Marking requirements.

The declaration of conformity/declaration of performance is in turn supported by a technical file which includes the following information:

- technical description;
- drawings, circuit diagrams and photos;
- bill of materials;
- specification and, where applicable, declarations of conformity for the critical components and materials used;
- details of any design calculations;
- test reports and/or assessments; and,
- instructions.

Technical documentation can be made available in paper or electronic format and must be available for a period of up to 10 years after the manufacture of the last unit.

6.6.3 Optical safety (photo-biological effects)

Lighting products have to comply with BS EN 62471, which contains requirements, including emission limits and a framework for classification, for all artificial lighting products that the eye and skin might be exposed to.

Exposure limit values published by the International Commission on Non-Ionizing Radiation Protection have been incorporated into European legislation (Directive 2006/25/EC, the Artificial Optical Radiation Directive). In the UK, this Directive has been implemented through the Control of Artificial Optical Radiation at Work Regulations 2010, which applies to exposure of workers to all artificial light sources including LEDs.

These regulations require employers to protect the eyes and skin of workers from exposure to hazardous sources of artificial optical radiation. This relates mainly to high-energy short-wavelength light and ultraviolet radiation. At very high intensities, blue spectrum light (short-wavelength 400 - 500 nm) may cause damage to humans if they are exposed to such lighting over a period of time and at, or within, a given distance.

a) Risk groups

Risk groups were developed to provide guidance on the potential optical safety hazard. These are a combination of source power and exposure time, and are divided into four risk groups for luminaires and other light sources:

- Risk group 0 (exempt; completely safe)
- Risk group 1 (low)
- Risk group 2 (moderate)
- Risk group 3 (high)

Lighting products intended for use in General Lighting Service should be assessed at the location where the illuminance is 500 lux. All other sources should be assigned to the relevant risk group based on an assessment 20 cm between an unprotected eye and the light source. However, the actual measurement does not need to be carried out at 20 cm, provided the assessment can be related to the exposure that would exist at 20 cm. PD IEC/TR 62778 *Application of IEC 62471 for the assessment of blue light hazard to light sources and luminaires* may also need to be taken into account.

b) Flicker effects and stroboscopic flicker effects

Flicker is a rapid and repeated change in light brightness, defined in terms of modulation or frequency. There are two types of flicker. The first is temporal, which is visible at frequencies up to about 100 Hz. The second is spatial, which can be perceived at frequencies up to and exceeding 2 kHz when the observer is moving relative to the source or when the head is moved quickly. Flicker may adversely affect health and wellbeing. The temporal variation may trigger seizures. Invisible flicker may cause headaches and other non-specific health effects. The human eye has sensitivity to flicker which varies with frequency and whether the source of flicker is viewed directly or in the peripheral field. The figure below shows the combinations of frequency and modulation that may give rise to problems. It is also important to recognise that high-frequency flicker may cause moving machinery to appear stationary due to the strobe-effect.

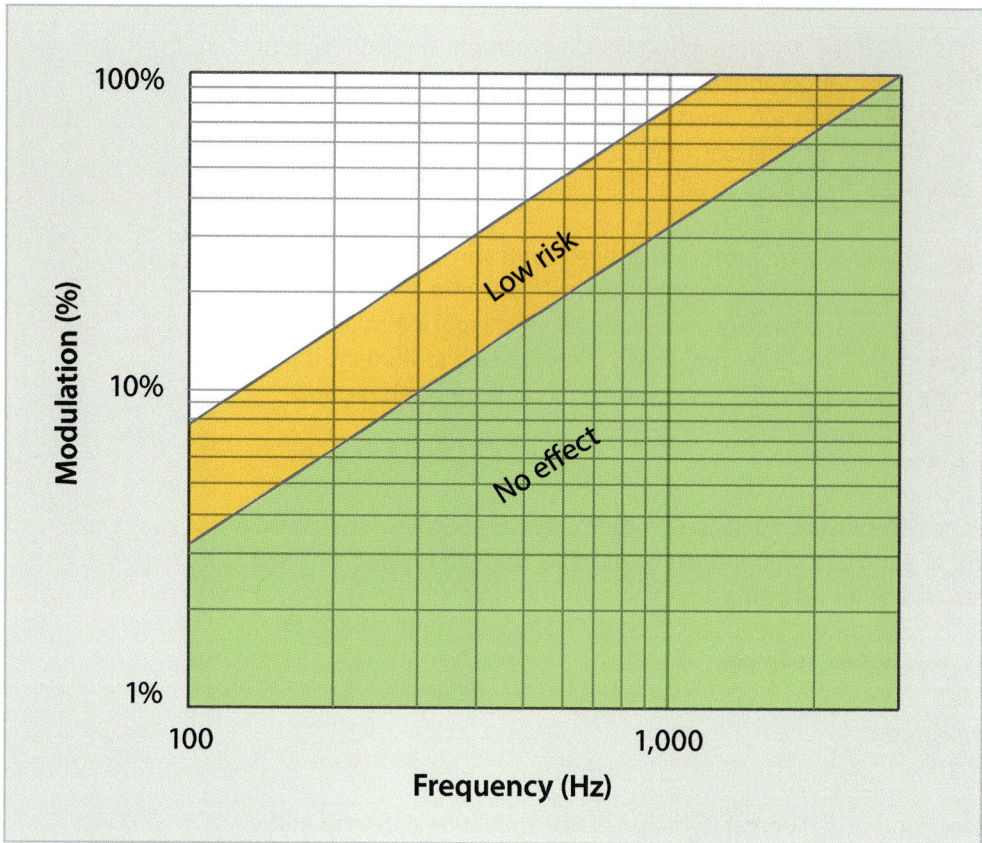

c) Infrared radiation (IR)

Generally, unlike light sources such as halogen lamps, energy efficient light sources do not emit significant IR radiation (unless specifically designed to do so) and the IR radiation is not powerful enough to pose any risks to humans.

d) Ultraviolet radiation (UV)

UV-driven phosphors may be used on some energy efficient light sources to create white light. However, at the distances at which these would be used in exterior lighting applications, these are unlikely to be an issue.

e) Blue light hazard

Blue light hazard (BLH) is defined as the potential for retinal injury due to high-energy short-wavelength light. At very high intensities, blue light (short-wavelength 400-500 nm) can lead to irreversible eye damage. For an injurious effect to occur from blue light hazard, three risk factors are critical:

- the amount of blue light contained within the total spectrum of the light source;
- the amount of light that can be coupled into the eye to expose the retina, determined by the radiance of the light source (at higher radiance, more photons are likely to hit photopigments and cause damage); and,
- the duration of exposure (at longer exposure, effects increase steadily) and cumulative duration of repeated exposures.

There is a potential safety hazard with risk of possible injury to the eye through excessive light intensity that may be emitted by an energy efficient light source at close distances.

6.6.4 Managing health and safety considerations

Responsible lighting manufacturers issue reliable and auditable information on the safe use of their devices. Many manufacturers undertake their own additional due diligence measures in order to show that they have considered and addressed optical safety hazards such that eye damage is unlikely to result through the responsible use of their products.

Lighting designers, installers and maintainers should work to ensure health and safety risks are identified and managed as appropriate for the lifecycle of the exterior lighting application, including eliminating unnecessary risks to public or worker health and safety.

Design brief in context

7.1　Introduction

The provision of good quality public lighting is an extremely important part of any infrastructure design. It ranges from lighting roads, paths and bridges to car parks, squares and amenities. It is essential for work, travel, and social activities; it can improve road safety, help reduce the fear of crime, aid regeneration and increase sustainability. However, its quality and performance rely on appropriate specification, design and planning, in addition to asset management and maintenance regimes for new and existing lighting stock.

NOTE: Maintenance routines should be considered from the outset of the design stage.

7.2　The initial design process

7.2.1　Designer and client communication

The design process starts with communication between the client and the designer(s)/ design team responsible for the lighting scheme (see Section 7.2.3 below) and its progress relies on the channels of communication between all the parties remaining open. It is vital to gain a full understanding of the client's needs and expectations.

Ensuring that the designer and client use a common language is important in order to avoid misunderstandings. Documentation should include a glossary where necessary to ensure comprehension. The designer should maintain a design checklist based on the client's requirements.

It is crucial to establish the following at the start of the process:

- The client's requirements.
- The nature and skills of the design team.
- A record and analysis of all proposed development and master plans.
- Legislation and standards.
- Technical information.
- Access to infrastructure and asset information.
- Ground rules and lines of responsibility.
- Existing problems.

7.2.2　The client's requirements

The client's requirements may include technical issues, development plans, compliance, environmental targets and policies. This may already also include a formal area, town or region-wide lighting strategy. The client may also be working within a wider policy framework and will require risk assessments at various stages.

While there is no statutory requirement to provide exterior lighting, there is great deal of legislation that either directly or indirectly imposes related powers and duties on asset owners such as local authorities, including a general duty of care for the public and road users. They are also subject to mandatory schemes such as CRCEE, which is aimed at improving energy efficiency and cutting emissions in large public and private sector organisations. In addition, there are likely to be budgetary rules and constraints, as well as the need to provide best value for money.

▼ **Figure 7.1** Example of footpath lighting (courtesy of CU Phosco)

7.2.3 The nature and skills of the design team

The design team may include internal or external lighting designer(s) who should be professionally competent for the tasks required – this competence is gained through a mix of experience, formal training and continuing professional development (see 1.4 Capability and competence).

The size and make-up of the design team will vary depending on the asset owner. Transport for London, for example, would expect a multi-disciplinary team to include specialists – including architects, modal (transport) experts, highway, traffic and maintenance engineers, a CDM Principal Designer, lighting/electrical engineer and stakeholders, including local authorities, the emergency services and security advisers – all led by a project manager.

7.2.4 Records and analysis of all proposed development and master plans

Before any design progresses, it is important that all relevant stakeholders such as agencies, service providers and local authorities are consulted on programmed works and planned development that might include or affect the lighting, including new building work. This will include the lighting engineer, local authority master planners, landscape architects, police and Highways England (or another regional highways authority).

7.2.5 Legislation and standards

Local, regional and national regulations, standards and policies should be followed. These include but are not restricted to:

- British and/or European standard recommendations for lighting parameters for roads, footways, cycle paths, conflict areas, external work areas, car parks, etc.;
- environmental zones lighting requirements;
- the need to limit or eliminate light pollution and intrusion;
- local authority planning requirements;
- heritage and conservation issues, including those relating to architectural and national monument sites, and protection of the countryside;
- the effects on flora and fauna;
- electrical installation considerations;
- maintenance plans;
- lighting controls considerations;
- columns/structures, including existing polices on height and spacing; and,
- the need to future-proof new lighting installations.

Typically, a local authority will define the road hierarchy in its area, including new buildings or infrastructure developments. Any lighting design will need to conform to standards for performance and design procedures specified by the asset owner, such as BS EN 13201, which details photometric standards for classes of road, and BS 5489-1, which gives recommendations on general principles of lighting for all types of highways and thoroughfares including those specifically for pedestrians and cyclists. This standard considers the design of lighting for all types of highways and public.

The asset owner may have its own additional lighting standards and requirements regarding schools, hospitals, conservation areas and areas of natural beauty. Early consultation should be directed at:

- eliminating confusion with air or water navigation lights or railway signals and ensuring the safe operation of other services;
- identifying the most appropriate lighting for locations in rural, environmentally sensitive areas and conservation areas; and,
- agreeing the use of any heritage or non-maintenance standard equipment.

The effects of artificial lighting on the ecology of an area must also be considered. Flora and fauna can be directly and indirectly affected due to the provision of night time lighting. Sleeping and feeding habits can change for birds, insects and mammals, which can lead to early exhaustion of food resources. Plants and trees are also vulnerable to the effects of artificial night time lighting where their pollinators (insects and birds) only come out at night. Plant growth can be affected with the plants reaching maturity before they should and then succumbing to climatic changes, such as late frosts that would have been avoided if the growth rate was normal. This could lead to reduced cover and food for insects.

The practical reasons for introducing LED lighting include greater energy efficiency, reduced maintenance, and reduction in energy use, lower emissions and potentially reduced light pollution, and improved safety. Local authorities are under increasing pressure to save energy, cut emissions and make cost savings, often in line with legislation. The designer is best placed to ensure that lighting solutions are efficient at an early stage in the project.

The advantages of any technology should not blind designers or clients to other social and aesthetic considerations. In particular, any lighting design should not overlook the quality of the lighting in an area or the requirements for colour rendition and colour temperature, which differ between, say, those for a loading yard and a suburban side street; a highway and residential area.

The guiding principle should be: the right light in the right place at the right time. To this could be added: 'under the right control.'

7.2.6 Technical information

The specifier, installer and end-user all require technical information regarding cabling, columns, luminaires, and controls, in order to support an informed decision on products and to maintain their asset management records system.

Much of the technical information required for any development proposals should be in the form of a specification spreadsheet/document that includes criteria in Table 7.1.

▼ **Table 7.1** Sample technical criteria for a specification spreadsheet/document.

Consideration	Sample criteria
Lighting parameters	<ul><li>Luminaire and light source efficacy</li><li>Colour temperature of the lamp/LED</li><li>Colour Rendering Index (CRI)</li></ul>
Control/driver	<ul><li>Control gear type: electronic/magnetic</li><li>Driver current</li><li>Requirement for control systems/functionality expectations</li></ul>
Power/cabling	<ul><li>Cabling type: internal and external</li><li>Supply type: 3-phase/single-phase</li><li>Earthing arrangements</li></ul>
Physical/structure	<ul><li>Column materials</li><li>Column installation method: flange plate or direct</li><li>IP and IK ratings</li></ul>

NOTE: Annex A provides a Good Practice Specification Template for energy efficient exterior lighting systems.

7.2.7 Access to infrastructure and asset information

Access to all relevant information on existing infrastructure, stock (including its condition) and maintenance programmes should be established and relevant records obtained.

7.2.8 Ground rules and lines of responsibility

Rules and responsibilities should be established for maintaining records and adhering to specifications. This includes the need for the asset management team to keep detailed records for all lighting stock – particularly where longer term warranties have been accepted or negotiated.

7.2.9 Existing problems

Where existing lighting will be upgraded, it is important to have a clear understanding of any existing problems, such as:

- Which areas are currently under/over lit?
- What proportion of the existing stock is defective and how much could be retained?
- How much energy could be saved through the addition of controls?
- Will decorative and heritage lighting be retained?

It is also important to manage expectations. If the proposal in the upgrade or project area is to retrofit LED luminaires, for example, the client will need to know that this is not the same as a like-for-like replacement. LED luminaires have a very different optical output from traditional high pressure or low pressure discharge lamp luminaires. Improved control of the light in one area might lead to lower light levels in surrounding areas.

Where the project is an upgrade of existing lighting, initial questions must be asked including:

- What ambient lighting level is present?
- Is there a need for additional lighting?
- Will the existing infrastructure (cabling and electrical supplies) and columns support the new lighting?
- Is there a change to area usage/classification?

Lighting designers also have to consider how the new lighting will interact with the existing environment, including other street furniture.

7.3 Lighting design options, modelling and evaluation

It is often necessary to model lighting scenarios and/or test them on site – perhaps as part of the stakeholder consultation exercise. Modelling software can show whether and how a luminaire will meet the lighting classification for a given site. In some cases it may also be necessary to model adjacent areas. For example, in a residential area, to show how the new lighting scheme will affect neighbouring properties. While street lighting is not there to light gardens, residents may be so accustomed to light spill that they would feel less secure if it were to disappear.

7.4 Site trials

LED luminaires vary in quality and design, and their effects may be difficult to visualise. Although software, product data sheets, manufacturers' websites and brochures are all useful sources of information, they cannot compare to visiting an installation at a manufacturer's test site, a neighbouring local authority or another asset owner site. Where possible, site visits should be arranged for key stakeholders, especially those whose expertise is not technical. Issues such as light spill into buildings adjacent to a lit highway can be modelled and displayed with a high degree of accuracy but can sometimes be easier to see on site and consequently to mitigate. It should not simply be assumed that a new type of luminaire will provide distribution and light control that is at least as good as the original.

A pilot scheme may be necessary to test both the new lighting and residents' reactions, and budgets or political considerations may dictate that, for example, city centre lighting is prioritised over traffic routes or residential. These and similar considerations will need to form part of any feasibility study.

In many cases where new technology, including LED lighting, is to be used it may be necessary to prepare a trial site where the lighting can be monitored to observe:

- lighting level and depreciation over a period of time;
- energy use and savings;
- material resilience to weather conditions and vandalism; and,
- residents' reactions to lighting levels, colour rendering and colour temperature.

The results of these trials should be used in feasibility studies. These will help key stakeholders reach informed decisions on issues such as where and when to install the technology on a larger scale in order to, for example, increase the use of city centre amenities. It can also help influence the setting of priorities such as lighting in areas with high crime rates where CCTV may be installed or to improve the image of the city centre and encourage tourism.

7.5   Construction, structural, civil works

The need to consult agencies, service providers and local authorities on development plans is especially acute in relation to brown field sites, for example, where there may be changes to the classification of an area, road widening, proposals for new amenities such as schools, nurseries and nursing homes, all of which will dictate the overall lighting strategy of the project.

Consultation will minimise any abortive and costly work and may reveal the need for additional planning. For example, how will the columns be installed? Root mounted or flange plate? Both methods are acceptable but have different requirements for the civil works: root-mounted requires a smaller area of excavation but can be deep depending on the column height, bracket length (if any) and luminaire weight; flange plate-mounted requires a larger area for the mass concrete foundation and may conflict with existing utilities.

In addition, the civil works associated with the lighting will have an environmental impact. Trenching at the installation stage will require excavation machinery, the disposal of excavated materials and the supply of new backfill material. The contractor's plant will add to the overall carbon emissions.

Among the considerations are:

- the location and nature of existing and proposed utilities;
- the presence of available power supplies;
- whether the electrical supply is via a Distribution Network Operator (DNO)/ Independent Distribution Network Operator (IDNO) connection or dedicated cabled network;
- whether cables laid directly in the ground or in a ducted system requiring only minor excavations to replace any damaged cable;
- clearance from the edge of the carriageway to minimise collisions by vehicles;
- whether footpaths are free from obstructions, including columns and/or foundations;
- visibility splays to ensure visibility of vehicles emerging from access, dependent on the length of road and normal road speed of traffic on the road outside the access;
- junctions, driveways and road geometry, including bends in the road;
- visual impact of the lighting on residents and other users of the public space;
- ease of access for maintenance; and,
- structural condition of the lighting column and brackets.

7.6   Lighting column retention/replacement

Asset owners can often make substantial savings by converting or replacing luminaires while retaining columns, and, of necessity, keeping their height and the spacing the same. In some cases, this may be to preserve the character of an area.

Considerations for lighting retention/replacement include evaluation of:

- the structural integrity of the column and any brackets;
- the differences in weight and wind area between existing and replacement luminaires (NB: LED luminaires can be substantially heavier than high pressure or low discharge lamp luminaires with remote gear, for instance);
- the adequacy, condition and nature of the existing cabling and circuit protection devices;

- the column spacing as this may either affect the choice of luminaire optic setting and may necessitate the use of multiple optic settings within the design, or require the replacement of sound columns;
- the asset owner's aims, such as the reduction of street clutter or a desire for increase accessibility; and,
- whether any retained columns will survive the predicted lifetime of the new luminaires, say up to 20 years.

7.7 Electrical power supply

The electrical supply can be single-phase or 3-phase; single-phase, two- or three- core cable utilises 230 V and is considered safer than a 400 V, 3-phase system. There are two main methods of providing electrical power to the lighting network:

- DNO/IDNO connection. This has the advantage of potentially minimising the ex-cavations required; only a relatively small joint hole is needed between the DNO/IDNO low voltage network and the column. A disadvantage is that only DNO-/IDNO-approved organisations or individuals are allowed to identify and repair faults on the LV cable, which can cause delays.
- A private cabled network. This is where a dedicated cabling system is used to supply the lighting network. This has one DNO/IDNO supply into a feeder pillar and the cable network feeds the lighting points. This network can have cables laid directly in the ground or in a ducted system and the asset owner is able to carry out any maintenance without reference to the DNO/IDNO. However, the capital cost may be higher as excavation across the whole length of the network is required.

The electrical design for the lighting systems has to consider the following points:

- circuit load: this includes inrush current, i.e. the combined load on each circuit.
- circuit length: the length of the circuit from feeder pillar to the end of circuit, including the cable length in each lighting column.
- cable type: steel wire armoured (SWA), cross-linked polyethylene (XLPE), split concentric, etc.
- cable cross-sectional area: $10mm^2$, $16mm^2$, etc.
- earth loop impedance: this is relevant at the feeder pillar supply and at the end of the circuit.
- earthing type: separate earth core in the cable and/or the cable armour.
- circuit breaker or fuse rating: ratings linked to circuit load and circuit disconnection time.
- voltage drop: the voltage loss across the circuit length.
- circuit disconnection time: the time taken for the circuit to disconnect under fault conditions.
- capital cost: determined by all of the above.

7.8 Lighting controls

Lighting designs need to be future-proofed as far as practicable, to be ready, for example, to incorporate additional services such as WiFi or CCTV where the column and cable design must take into account the additional weight, wind area and load of the additional equipment. Luminaires should be specified with lighting controls in mind. Lighting controls are used increasingly used to save energy and make maintenance more efficient, and range from photo-electronic sensors or pre-programmed profiles stored in a luminaire's electronics through to more complex CMS or telemanagement systems.

A CMS enables lighting to be trimmed, dimmed, varied/adapted or switched off based on criteria such as traffic flow or the changing requirements of a night-time economy. Adaptive lighting controls allow asset owners to respond quickly to requests for more light in a given area, for example, in an emergency, or simply to save electricity. A CMS also gives the asset owner greater control over lighting stock, making it easier to identify, predict or respond to maintenance requirements and to fix faults.

A risk assessment will be required in relation to part-night lighting or switch-offs; some exemptions may be required by for road safety purposes or at the request of the police.

Once installed the system should enable the asset owner to:

- identify lighting system problems in advance;
- identify failures;
- potentially predict failures;
- inform maintenance routines;
- manage the power consumption; and,
- remote-control system on/off times and lighting levels.

7.9 Maintenance plans

The maintenance plan for any external lighting installation should be understood before any works start. This will include a maintenance risk assessment that covers issues such as whether luminaire asset maintenance will involve major road closures and whether the nature of the site calls for individual lamp or complete module replacement. Factors such as environmental zones, traffic quantity, column heights, burning hours and the category of the road should be taken into account, and lighting asset cleaning maintenance regimes should be developed.

The maintenance plan should also include provision for planned maintenance such as group light source replacement, depending on burning hours and lamp lumen output depreciation over time, or LED module or LED luminaire replacement when the luminaire lumen output falls below the designed level.

Practical considerations when specifying products include:

- their weight: products above a certain weight may require two-person handling or specialist installation or maintenance;
- access for maintenance: some products offer simple tool or tool-less maintenance and plug and socket module and component replacement; and,
- cleaning: while LED luminaires should, in theory, only need to be inspected, for example, every six years, in reality they may fail for a variety of reasons and will certainly require cleaning.

NOTE: Whether the source is HID, fluorescent or LED, a dirty luminaire will eventually fail to meet lighting level recommendations for that particular public space.

7.10 Lighting inventory and asset management

Asset owners need detailed and current inventories of existing lighting in order to evaluate assets, gain funds to improve them and to deliver best value. For instance, a detailed street lighting inventory enables an asset owner both to allocate funds and to provide the information that others – such as the government for local authorities – use for funding allocations.

Many asset owners already realise the importance of asset management and use dedicated software designed to help manage exterior lighting systems. Ideally, the entire system should be on the asset management system, including electrical supply, column details, position, luminaire types, and any information relating to the use of the space that may help determine future lighting requirements or upgrades.

Detailed information on maintenance management on highways can be found in the latest edition of the *Well-lit Highways – Code of Practice for Highway Lighting Management*. It includes a sample list of attributes for each asset that should form the core basis of any local authority's asset record system. ILP TR22 *Managing a Vital Asset – Lighting Supports* also gives information on managing inventory, condition assessment, condition indicators, and lifecycle forward planning of a local authority's lighting support stock.

7.11 Greenfield and brownfield sites

Lighting design for greenfield sites is relatively straightforward compared to designing for brownfield sites:

- the lighting category and environmental zone will have been determined;
- the design team and client will have reached decisions on column heights, style, source and luminaire style;
- the road designer and landscape architect will have worked on the road require-ments – lane and footpath widths – and a range of public lighting requirements;
- the lighting designer will use modelling software to determine the inter-column spacing on the mainline and the column configuration at the conflict areas (for example, junctions, roundabouts and pedestrian crossing areas)
- the lighting designer will plot the necessary network column positions and execute the electrical design; and,
- power supplies in greenfield areas will be coordinated with the service provider and connections agreed.

Brownfield sites, however, present very different scenarios and challenges, some of which have been covered elsewhere in this Section, such as existing utilities, civil works and existing column conditions. The lighting design may only concentrate on upgrading of the lighting inventory, for example, when it is near the end of its life, when it has become energy-intensive, if it is unsafe or if it fails to comply with the asset owner's master plan or with the vision for the future.

If budgetary constraints do not allow a full upgrade of the lighting columns and the electrical network, the lighting designer may have to consider only replacing luminaires. In that case, the designer needs to follow these steps:

- Check column structural integrity to ensure that the columns, including projection brackets (if any) are capable of carrying the new luminaires safely. The residual life of the columns should be calculated to ensure that the whole life costing can be accurately calculated and to avoid the cost of installation works being carried out twice in a short space of time.
- Check luminaire mounting criteria including where older lighting networks may have luminaires where heavy magnetic (wire wound) ballasts are remote from the luminaire and are housed in the column; and to take into account the fact that some LED luminaires are considerably heavier and may be larger than the old discharge lamp luminaires and present issues of increased weight and wind area.
- Check the electrical cable network integrity to ensure that the cable complies with the relevant standards. BS 7671 *Requirements for Electrical Installations* covers assessment of general characteristics as well as providing a section on electrical testing requirements.

© The Institution of Engineering and Technology

NOTE: Where only luminaires are being replaced, a minor works certification is required.

- Reassess circuit protection (circuit breakers/fuses) to ensure compliance with the requirements of BS 7671 in terms of maximum impedance (Zs), load, and disconnection time.
- Check that the electrical supply will cope with the proposed load, which may necessitate discussion with the DNO/IDNO to ensure that the load demand is available.

Only after these items have been checked and certified as compliant should the lighting design commence.

Additional issues in this kind of scenario should also be captured within the DRA, including:

- whether the position, spacing and height of lighting columns are fixed (i.e. they are not being replaced as part of a new design), as if so the lighting designer can face various obstacles to attaining the required average lighting levels and uniformities;
- whether options exist to minimise the number of luminaire optic settings and wattages proposed in order to reduce installation and maintenance issues, such as having to ensure that each defined optic setting is installed on the correct column and replaced correctly during the installation works;
- whether it is possible to achieve the required minimum lighting level in all areas; in some cases, certain areas may need to be over-lit or the client may need to agree a relaxation of levels/uniformities.

Lighting system component specification and selection

8.1 Introduction

This Section covers the specification and selection of components for an exterior lighting installation, which may be made up of the following elements:

- luminaire;
- lighting column or structure;
- cable network;
- feeder pillar;
- ducting system;
- over-voltage and surge protection;
- over-current and circuit protection;
- cut-out; and may also include,
- additional electrical supplies in lighting columns.

8.2 Selection process

The selection of the various components of a lighting system must go through a logical process to ensure that the equipment has met the specifications from the local authorities or clients who will adopt and maintain them for the coming years. The selection process must consider the following items:

- materials used in construction;
- longevity of the components;
- compliance with safety and environmental requirements including RoHS, WEEE, BS EN 40;
- whole-life costing; and,
- uniformity in design.

The lighting designer should follow a design checklist to ensure that the equipment specified meets the requirements of the client and/or the asset owner. It should be an easy-to-follow log of all decisions made during the design process. This checklist will ensure that the designer has considered all of the requirements and can 'tick off' the items as the design progresses. Where equivalent or possibly non-compliant materials are proposed, a statement (which may be a mitigation statement) can be prepared to detail why the material or item has been selected and can be issued to the client for discussion/approval.

Where items are included on the checklist it is important that the required standards are identified to ensure that the designer is working with the most up to date revisions and amendments/addenda.

The lighting layout should include:

- a schedule of DNO/IDNO connections, disconnections and transfer, plus details of any private cables and their feeds;
- all existing and planned lighting point locations in the immediate and surrounding areas covered by the lighting design;
- a schedule of equipment, and parameters used in the design, including lighting class and maintenance factors; and,
- calculations indicating the surround ratio, maximum, average and minimum levels of luminance or illuminance and uniformity factor to demonstrate compliance with required standards and/or guidance for lighting parameters.

8.3 Luminaires

8.3.1 General luminaire selection criteria

Luminaires chosen for external public spaces need to perform the lighting task required for that particular space. They should be robust and reliable, require the minimum of maintenance and be recyclable at the end of their life. All luminaires should conform to BS EN 60598 and be CE marked with supporting data.

Luminaires vary considerably in size and style, application, construction materials, performance and cost, mounting adaptability, ease of maintenance and recyclability. Luminaires should adhere to the requirements for minimising:

- light pollution;
- light intrusion;
- glare; and
- energy use and emissions.

The choice of luminaire can depend on a number of factors including style and aesthetics, construction materials, RAL colour finish, ease of maintenance, performance, adaptability, recycling properties and price – but all should be fit for purpose.

Other considerations include the IP and IK ratings required for the application of the lighting unit. A minimum of IP65 ingress protection would normally be required, with in-ground up-lighters or marine environments requiring IP67 and with, commonly, an impact protection class of IK10.

The method of electrical supply must also be considered and electrical safety characteristics determined to ensure the safety of the public and maintenance operatives. DRAs should be used to highlight the requirements and regulations associated with the design and to record if any special requirements relate to a particular site.

8.3.2 Luminaire selection for particular applications

Any products that are to be used on a local authority unmetered supply should have Elexon UMSUG charge codes available.

Luminaires specified for roads, footways or cycle paths should meet the performance requirements for that application including the delivery of the appropriate amount of light in the correct place and using products that meet any glare limits placed on that particular category of road.

Luminaires for car parks or open areas should be selected so that areas can be lit but not over-lit using products that emit a minimum of upward light.

Tunnel lighting is more complicated than road or architectural lighting, mainly because it involves lighting the tunnel in stages to compensate for the time the eye takes to adjust from light to dark and back. These are extreme environments and specialist advice should be sought before specifying products for road tunnels.

▼ **Figure 8.1** Example of tunnel lighting (courtesy of Indo)

Products used for amenity lighting may need to be robust and vandal-resistant. Visually appealing columns and luminaires that form part of the urban landscape are a common requirement in the redevelopment of many of our modern towns and city centres. Urban landscape lighting developments may include lit street furniture or in-ground light sources or bespoke combined integrated column and luminaires. Luminaires with the correct performance characteristics for the space to be lit should be selected. Amenity lighting will, in most cases, be mounted on lower columns/structures or as in-ground units, and as such may need to be specified to a higher standard than lighting at greater mounting heights. The designer must also consider the likelihood of a luminaire being accessed by members of the public, by accident or through vandalism.

8.3.3 Luminaire certification

The luminaire supplier must provide certification that the minimum standards recommended in the Good Practice Specification Template (www.theiet.org/exterior-lighting) are achieved for each proposed luminaire type.

8.4 Lighting columns

8.4.1 General lighting column selection criteria

The variety and type of lighting columns for different applications is immense. Columns can range from simple off-the-shelf standard design to intricate bespoke manufacture. Typical applications would include road lighting; amenity lighting; floodlighting; stadium lighting and high mast applications. Other applications for columns may include catenary systems and seasonal (for example, Christmas) decorations. Column design and application must take into account geographical location, terrain, topography, altitude, bracket weight (if any), additional fixings and luminaire weight and wind area.

Designs include stepped, tapered, mid-hinge, base-hinged, heritage and passively safe columns – the latter for high risk road lighting installations. There is a multitude of bracket types available and winch systems for hinged column or high mast sites. Columns can be installed rooted or surface-mounted, subject to design calculations and site conditions.

There is a movement towards integrating other services or equipment into street lighting columns to assist in reducing street clutter and the number of different columns required for different applications. Current examples include integrating traffic signals, WiFi and CCTV cameras within or onto lighting columns.

The condition and type of the existing column stock should be recorded on the asset management system. If retrofitting luminaires on to existing lighting stock, suitable checks should be made on the light distribution, weight and wind area calculations. Where columns are likely to support additional loads such as traffic signs, flower baskets, banners/flags and Christmas decorations, the column should be designed to carry the additional load.

8.4.2 Lighting column design and construction

Columns are usually made from steel or aluminium and are available with a variety of protective coatings, though the majority of lighting columns are hot-dipped galvanized.

Locations such as those near to overhead power lines have particular requirements. High mast installations are normally bespoke and often have civil engineering involvement in the design of the very deep foundations that are required. Spigot entry sizes of columns and brackets should always be checked against the spigot entry sizes of any proposed luminaires or other equipment.

BS EN 40-3-2 *Lighting columns. Design and Verification* gives details of the design of lighting columns and provides information for use in the design and application of exterior lighting systems.

Passively safe columns may be required on major urban roads whereas in areas where speeds are low, such as housing estates, there is little if any advantage to using them and the risk to pedestrians may be higher when compared to conventional columns. ILP TR30 *Passive Safety: Guidance on the Implementation of Passively Safe Lighting Columns and Signposts* provides guidance to highway lighting designers on identifying locations, carrying out risk assessments and selecting apparatus in accordance with BS EN 12767 and BS EN 40.

PD 6547: *Guidance on the use of BS EN 40-3-1 and BS EN 40-3-3* also provides information on the key standards for the design and verification of lighting columns.

8.5   Electrical installation considerations

8.5.1 Regulations and standards

The Electricity at Work Regulations require that full details of all electrical equipment are recorded and made available to those operating and maintaining it. All installations should conform to BS 7671. The New Roads and Street Works Act requires that the locations of electrical services and circuits are also recorded and must be made available to any 'statutory undertaker' planning to excavate the highway.

8.5.2 Feeder pillars

Feeder pillars are mechanical enclosures for the electrical control gear for distribution to lighting circuits. They often contain the main electrical supply from the DNO/IDNO and the circuit protection (circuit breakers/fuses). Other control gear such as contactors, timers, PECUs and CMS systems may be contained inside the feeder pillar enclosure. All feeder pillars must be lockable, preferably with a locking system with anti-vandal/tamper locks. Feeder pillars are usually constructed from galvanised sheet steel or, if the area dictates that the pillars should be decorative, stainless steel.

Feeder pillars may be manufactured to BS EN 61439 *Low-voltage switchgear and control gear assemblies. Assemblies for power distribution in public networks.*

8.5.3 Ducting system

Lighting ducting systems protect cable and ease the replacement of the cable in instances where faults have occurred. Where a ducted system is used in street lighting, it should conform to BS EN 61386-24 *Conduit systems for cable management. Particular requirements. Conduit systems buried underground*, and – in public areas – the NJUG *Guidelines on the Positioning and Colour Coding of Underground Utilities' Apparatus.*

Lighting ducting should be coloured for ease identification during excavations. Care must be taken to ensure that the colour specified matches that of the NJUG National Colour Coding System. Street lighting ducting in England and Wales is coloured orange whereas in Scotland and Northern Ireland it is coloured purple.

8.5.4 Overvoltage and surge protection

When a new structure or system is installed, consideration has to be given to the possible effects caused by either a direct or indirect lightning strike. Consideration should also be given for the protection of electrical installations against transient overvoltages of atmospheric origin transmitted by the supply distribution system or due to switching overvoltages generated by the equipment within the installation.

To achieve these goals it is recommended that the design, installation and testing of the lightning and surge protection systems be carried out by a suitably qualified specialist. This may entail the installation of surge protection devices, subject to appropriate risk assessment. This should take into account the requirements of other disciplines working on the project, especially with regard to the different types of earthing systems that may be specified.

Further information can be found in the BS EN 62305 *Protection against lightning*, BS 7671 (sections 443 and 534), and the BEAMA *Guide to Surge Protection Devices.*

8.5.5 Over-current and circuit protection

All lighting circuits must be protected by a circuit protective device, typically a cartridge fuse or a circuit breaker. The protective device is used to protect the cable from over-current during fault conditions.

The selection of the fuse type and rating must be carried out according to the circuit load including inrush currents, maximum Impedance (Zs) and required disconnection time. The disconnection time for highway street lighting distribution circuits is 5 seconds.

Although circuit breaker recommendations are widely available for discharge lamps, they can be difficult to find for LEDs, though some LED manufacturers publish inrush current/circuit breaker size and number of drivers per circuit breaker on their websites.

8.5.6 Cut-outs

Cut-outs are used as a termination point for the incoming cables from (and any outgoing cables to) the lighting system's electrical supply. They also contain a circuit protective device that is used to protect the luminaire when a fault occurs. Cut-outs may be used as a means of local isolation for the power supply to the luminaire.

Cut-outs must conform to BS 7654 *Specification for single-phase street lighting cut-out assemblies for low-voltage public electricity distribution systems.*

8.5.7 Additional electrical supplies within lighting columns

With the drive to reduce street-clutter, asset owners may look to multi-use or 'smart' poles where more than just lighting can be mounted on the lighting column. Items such as CCTV cameras, traffic lights, seasonal decorations and many other items (for example, potentially electric vehicle charging points) are being (or could be) added to or integrated within lighting columns. Each of these items relies on electrical power to function and adds weight.

Equipment must be manufactured, installed, maintained and operated in compliance with relevant European and UK product standards. Electrical installations intended for connection to the lighting system must comply with BS 7671 (additional guidance is provided within the IET Guidance Note series and the IET *Code of Practice for Electrical Safety Mangement*).

▼ **Figure 8.2** Summary of typical column dimensions for combined column installation of traffic signals and street lighting; measurements in millimetres (courtesy of TfL)

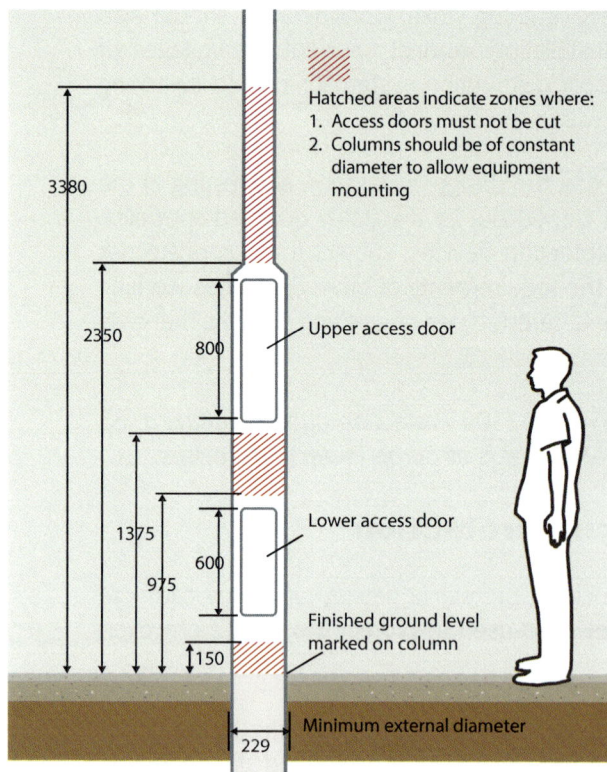

Notes:
- Circuit separation should be applied to ensure access by skilled staff.
- The planting depth shall be clearly marked on the columns to ensure signal heads and push buttons are positioned at the correct heights.
- Standard heights should be applied for signal equipment (e.g. pedestrian push / wait buttons), if these are to be installed. If not required, these restraints can be relaxed.
- Power supplies to a combined column installation pole should be from the same supply / phase.

8.6 Lighting controls

A number of manufacturers of lighting controls provide products that include stand-alone systems that use simple photocells, part night varying, programmable multi-level night varying CMS or pre-programmed constant light output systems.

A CMS allows the operator to choose exactly when to switch each individual light on or off or to vary the lighting level as required. CMS manufacturers offer a variety of systems of varying complexity. There is no standardisation of CMS at present, but there are developments that will enable systems to talk to each other. The specification for controls should include system integration if there are a number of control systems from different manufacturers already in use or if development work on an exterior lighting project requires new equipment to be linked to an existing CMS system.

CMS for use in Balancing and Settlement Code (BSC) unmetered supply systems need BSC approval.

More details are available, including on approved CMS, on the Elexon website – www. elexon.co.uk. The ILP TR27 *Code of Practice for Variable Light Levels for Highways* may also provide helpful information.

SECTION 9

Lighting scheme installation, commissioning, operation and maintenance

9.1 Introduction

Anyone carrying out installation, commissioning or maintenance works, including electrical inspection and testing should be competent to do so. For highway electrical installations, persons carrying out installation, commissioning or maintenance works, including inspection and testing should be registered to the Highway Electrical Registration Scheme (HERS) and hold an appropriate ECS HERS card. They should also have evidence of understanding and familiarity with this Guide.

9.2 Installation and initial verification

During and on completion of the installation, and before being put into service, new or modified installations should be inspected and tested to verify, initially, that the installation complies with BS 7671, the manufacturer' instructions and the designer's requirements. Records must be provided by the designer in order to ensure that this can be achieved. Particular attention should be paid to existing luminaires, with older lamp types, which have had control gear circuits modified to accept different lamps.

On completion, an Electrical Installation Certificate based on the information in Appendix 6 of BS 7671 together with relevant schedules of inspection and tests should be completed. This should be provided to the person ordering the work. All electrical works installed should be carried out to be in compliance with the requirements of BS 7671 and, as necessary, BS 5266 *Design for Emergency Lighting*. Care should be exercised to ensure that sensitive electronic equipment (for example, LED driver, surge protective device) is not damaged as a result of the tests being carried out.

Initial lighting levels should also be measured to ensure an appropriate base-line for subsequent periodic measurements.

9.3 Commissioning

Where commissioning is required, other than just functional testing, this should be carried out by suitably skilled persons competent to do so and in accordance with the manufacturer's instructions and as agreed with the asset owner.

9.4 Lighting system maintenance, inspection and testing

9.4.1 Routine maintenance

Cleaning of luminaires should be carried out in accordance with the manufacturer's, designer's and client's requirements. Cleaning should be carried out at pre-determined regular intervals, before lighting measurements are carried out or following a review of the results of lighting measurements. Other factors that may influence the cleaning frequency include the immediate environmental conditions and the IP rating of the luminaire itself. Care needs to be taken not to damage the light source through the cleaning process, particularly with LED products that contain remote phosphors.

9.4.2 Reactive maintenance

A logical approach to fault-finding should be adopted if a failure is not obvious. In the absence of any instructions to the contrary, items should be replaced on a like-for-like basis. Original light source characteristics should be ascertained to ensure that replacement components are as close to the original specification as possible, particularly in terms of the intensity and distribution of light, colour temperature and colour rendering.

Repairs, replacements or additions should leave the installation no less safe than originally installed, inspected and tested.

9.4.3 Lighting level testing

Periodic testing of the lighting levels should be carried out at regular intervals (in line with electrical inspection and testing):

- to ensure that the output of the luminaire is still performing within the declared limits from the designer/manufacturer taking into account the design requirements and failure mode (particularly of LEDs); and,
- to note any changes from the lighting design parameters, the original lighting measurements taken after initial commissioning and the previous periodic lighting measurement, if any.

Lighting levels can only be properly established by using an appropriate calibrated illuminance meter (or luminance meter as appropriate) and carrying out measurements. Illuminance measurements would need to be representative of the LED source(s) within the luminaire, i.e. for multiple LED sources within a luminaire, a single point illuminance measurement would be insufficient. Where colour rendering or colour appearance is considered important, a method of identifying unacceptable changes or variations should be agreed with the manufacturer or designer.

Periodic lighting testing should be carried out with records taken of any changes from the lighting design parameters, the original lighting measurements taken after initial commissioning and the previous recorded periodic lighting measurements.

9.4.4 Periodic inspection and testing of electrical condition

Periodic electrical inspection and testing on existing installations should be carried out in accordance with BS 7671 and associated guidance (for example, IET Guidance Note 3, *Inspection and Testing*). Upon completion of a periodic inspection with relevant test(s), an Electrical Installation Condition Report together with schedules of inspection and tests based on the information in Appendix 6 of BS 7671 should be completed (see also HEA standardized inspection and test certificate). Care should be exercised to ensure that sensitive electronic equipment (for example, LED driver, surge protective device) is not damaged as a result of the tests being carried out.

Routine checks (i.e. visual inspection with functional on/off checks) and electrical inspection and testing of LED lighting systems should be applied in accordance with the equipment manufacturer's recommendations and the designer's/client's requirements.

In the absence of any such recommendations or requirements, electrical periodic inspection and testing should be undertaken with the frequency recommended in IET Guidance Note 3.

9.5 Equipment management/disposal

Appropriate details of the equipment installed and modifications carried out should be recorded in order to ensure proper management of the asset. This may include, for example, details of LED driver settings where these vary from unit to unit. Redundant items and components should be disposed of in accordance with legislative requirements and where appropriate in accordance with the WEEE Regulations through an approved producer compliance scheme (for example, Lumicom).

Energy Efficient Exterior Lighting Systems – Good Practice Specification Template

A.1 Using the Good Practice Specification Template

This Good Practice Specification Template provides a basis for specification of exterior lighting systems and the evaluation of supplier offers for suitable solutions. It is designed to be used to set out and assess criteria so that alternative offers and lighting solutions can be assessed on a like-for-like basis.

The template has been designed to be broadly an output specification in terms of lighting equipment, and not to prevent the use of products that are part of a fast developing technology through the use of overly prescriptive input requirements. What is important is that the right type and amount of light is available for the required period of time (the lighting design life of the installation) and covers the required area in line with the appropriate standards.

NOTE: The specification template contains a significant number of parameters (aligned with relevant national and international standards) and so only occupationally competent individuals trained in the relevant IET Guidance should use and interpret this spreadsheet.

This Good Practice Specification Template is also available to download as an Excel spreadsheet from (www.theiet.org/exterior-lighting).

A.2 Good Practice Specification Template

	One worksheet per typical section/layout allowing multiple lighting solutions;
Notes	LED only rows highlighted green and start with 'LED';
	Road lighting only start with 'Road Lighting';
	Area Lighting highlighted blue and start with 'Area Only'

	Item	Value
To be Completed by Client/Specifier (Preferred Output for Road Geometry and existing column positions is that used by the usual proprietary software for lighting design)	**Sample Section Geometry for Comparison Purposes**	
	Road Lighting - Lane Width(s)	
	Road Lighting - Number of lanes	
	Road Lighting - Footway (Hard shoulder) Width(s)	
	Road Lighting - Central Reservation width	
	Road Lighting - Road Surface (Luminance)	
	Road Lighting - Setback (Kerb line to centre line of column)	
	AREA ONLY - Description of Area and Use *types of traffic etc.)	
	Bracket Projection	
	Tilt	
	Column spigot size - diameter x length (mm) (where existing columns used)	
	Column layout configuration (e.g. staggered, opposite)	
	Typical Spacing(s)	
	Mounting Height	
	Lighting Parameters	
	Road Lighting - Lighting Class	
	AREA Only - Lighting Levels - Average, Uniformity	
	Luminous Intensity Class (Glare)/'G Rating' (min) at zero tilt	
	Luminous Intensity Class (Glare)/'G Rating' (min) at scheme tilt (if different)	
	CRI (min)	
	Correlated Colour Temperature Range	
	S/P Ratio	
	Lighting Scheme Design Life (Operating Hrs)	
	Cleaning Frequency (Yrs)	
	Environmental Class (for depreciation)	
	IP Rating - Luminaire - Optical	
	IP Rating - Luminaire - Control gear	
	% Dim state if not 0 % dimmed (i.e. 100 % light output)	

Solution # (Repeat for additional Solutions)	
Luminaire/Lighting System for the Scheme above (at rated ambient temp 25 °C)	
Luminaire Manufacturer	
Model Name/No.	
LED Manufacturer & Model No.	
LED module replaceable? Y/N	
LED Driver replaceable? Y/N	
LED Photometric Code	

Item	Value
LED Rated Chromaticity Coordinates Initial	
CRI	
Correlated Colour Temperature (K)	
Luminous Intensity Class (Glare)/'G Rating' (min) at zero tilt	
Luminous Intensity Class (Glare)/'G Rating' (min) at scheme tilt (if different)	
Lumen output for luminaire at 36,000hrs (burning hours) for comparison - provide calculations	
Lumen output for luminaire at end of customer lighting scheme - provide calculations	
Rated Input Power (W)	
Circuit watts Including constant light output but otherwise Undimmed (Average over lighting scheme life)	
Power Factor - minimum power factor over lighting scheme life	
UMSUG code (where unmetered supplies used)	
Surge Protection - state type and protection provided	
Power Density Indicator (D_P or PDI) W/lx/m^2	
Annual Energy Consumption Indicator (D_E or AECI) Wh/m^2	
Luminaire Weight	
Luminaire Wind area (max)	
Luminaire spigot entry size	
Does the luminaire have the facility to be tilted (specify range or values)	
Photobiological risk group - provide test data (& where applicable calculations) for the public and also for maintenance personnel	
LED Rated median useful life (h) and associated maintenance factor (x)	
LED Rated abrupt failure value (%)	
Warranty Period - Lighting System including all components (yrs/hrs)	
Max Inrush Current (for fuse ratings/cable calcs/circuit protective devices)	
LED Relevant performance at 15 °C	
CE Marking DoC's (attach) LV + EMC + ErP + RoHS	
IP Rating - Luminaire - Optical Compartment	
IP Rating - Luminaire - Gear Compartment	
WEEE Producer Compliance Scheme Certification	
Manufacturer's instructions to be provided in English	

ANNEX B

Technology development for exterior lighting applications

B.1 Introduction

This Annex provides context for Section 3 and describes recent developments in exterior lighting technologies and the move towards a systems approach to lighting in general. Figure 3.1 shows the development of light sources and the increasing luminous efficacy over the last 75 years.

B.2 Flame-based sources

The original light sources revolved around fire which has a luminous efficacy of around 0.3 lm/W. There is still some outdoor lighting that uses gas mantles as the light emitter although these are more efficient than a naked flame at 1 to 2 lm/W.

The use of gas mantle lanterns tends to be restricted to heritage sites, for example, in London Parks and the Royal Estates. They are also still used in other European cities. Where still in use, the lanterns are often protected or listed under heritage planning laws and as such continue to be replaced like-for-like or refurbished. Other both new and traditional 'flambeaux' products are available that provide an architectural statement and contribute to the atmospheric lighting, although these are typically not primarily for illumination.

B.3 Incandescent

Electric lighting began with the introduction of incandescent lamps. In general, an electric current is passed through a filament in a vacuum, generating lots of heat and a little visible light. The vacuum keeps oxygen away from the filament, preventing it from oxidising and evaporating. Early versions used a carbon filament made from bamboo but later developments used tungsten filaments and replaced the vacuum with low pressure halogen gases to produce a brighter light with less degradation and with a longer lamp life. Despite development, modern tungsten halogen lamps still have efficiencies less than 5 %, typically having a luminous efficacy of 10-20 lm/W and a life of 1,000 or 2,000 hours.

B.4 Gas discharge

Gas discharge lighting is the most common light source in exterior lighting. There are several types of gas discharge lighting:

- fluorescent;
- mercury vapour;
- low pressure sodium;

- high pressure sodium;
- quartz metal halide; and,
- ceramic metal halide.

Gas discharge lamps do not have a filament and rely on the production of an arc between two electrodes to create light. The voltage required to create the arc in air would be impractical for lighting products and so an arc is created within a tube filled with inert gas. The arc tube is dosed with different metal compounds to promote the creation of the arc and to determine its colour. Mercury and sodium are the predominant metals used in lamps with mercury creating white light sources and sodium producing efficient orange/gold light sources. Some lamps, such as fluorescent lamps, use a phosphor compound coating on the inside of the arc tube or lamp envelope to convert ultraviolet light into green and red wavelengths which combine to give a white light.

To operate reliably, gas discharge lamps usually require control gear; either electromagnetic or electronic versions are available. Electromagnetic control gear usually includes an igniter which produces a series of high voltage pulses to initiate the arc, a ballast or choke which acts to restrict and control the current flow through the arc, and capacitors that neutralise the inductance created by the ballast. Electronic control gear creates the same effect often running the lamp more gently and so extending lamp life, as well as having improved electrical efficiency typically saving 10-15 % energy use relative to electromagnetic gear.

As gas discharge lighting has developed, the gas pressure inside the arc tubes has increased and this has improved the efficiency of the lamps. Small quantities of other metal salts are also added to tune the wavelengths of light emitted, improving colour temperature and colour rendering.

B.5   Fluorescent

Fluorescent tubes were one of the first low pressure mercury vapour gas discharge lamps developed at the start of the 20th century. The light produced is shortwave ultraviolet and so the inside of the glass envelope is coated with phosphor compounds selected to render a white light. The resulting light source is significantly more efficient than incandescent sources.

The disadvantage of early fluorescent lamps was the physical size of the straight tube required to generate sufficient light. Compact fluorescent lamps (CFL) were developed in the 1980s to reduce the size of the lamp. Reducing the diameter of the glass tube and folding the tube into meanders or loops also allows the lengths of the lamps to be reduced, meaning the source can be fitted into much smaller lanterns, reducing the overall costs. Various CFL solutions have been developed with both remote and integral control gear.

With the different variations of lamp, there is a wide range of luminous efficacies from 45 to 105 lm/W depending on the control gear type and fluorescent tube geometry. Luminous efficiencies are typically around 10 % and the lamp life is much improved at 6,000 to 15,000 hours.

B.6   Induction lighting

Electrodeless discharge lighting was first invented in the late 19th century. However, it wasn't until the 1960s that a practical induction lamp became a realistic light source.

There were further technological developments in the 1990s that brought the technology to the level at which it performs today. Some companies have tried to combine metal halide and fluorescent technology to further progress induction lighting but, at the time of writing, the resulting systems are large and relatively impractical for outdoor lighting.

The induction lamp is a fluorescent tube (a glass tube with the inside coated in phosphor compounds). The gases inside the tube are excited by a high frequency electromagnetic field generated around the lamp by the control gear. In this way, the electrodes can be omitted, improving the reliability of the lamp and increasing lamp life. However, this makes the control gear the critical component for reliability and the system life is dependent on the design and manufacturing of the control gear, which varies in quality and price. The life of the control gear will also depend on the ambient temperature of operation of the equipment. Luminous efficacy of induction lamps ranges from 65 to 87 lm/W, with lamp life from 80,000 to 100,000 hours.

While it has been available commercially for many years, induction lighting has not seen a significant adoption level for exterior lighting due to the high costs. More recently new companies have entered the market focusing on induction lighting, using the long lamp life and reduced maintenance costs as the main selling point. However, the large lamp sizes involved makes the bodies of the luminaires over twice the wind area of modern discharge lanterns, which in turn makes it difficult to retrofit induction lighting sources without overloading the existing lighting columns. Optical control of the light source is difficult due to the large lamp size and while this results in acceptable uniformity levels, light spill onto adjacent areas can reduce the system efficiency and limit energy savings.

B.7 Sodium vapour

There are two versions of sodium vapour lamps; low pressure sodium (abbreviated as LPS or SOX) produces a single wavelength of orange light and high-pressure sodium (HPS or SON), which has a paler, more golden light.

LPS is a pure sodium source and has a very high luminous efficacy compared to other high intensity discharge (HID) lamps. Like induction lighting the large lamp size of LPS makes the luminaires quite large compared to modern lanterns, with similar optical control and energy efficiency difficulties.

HPS lamps may include mercury or other metals to improve colour rendering but are less efficient as a source. However, they have a much smaller arc tube that allows the lantern's reflector to control the light output onto the ground more accurately. The improved optical control reduces wasted light and increases the system efficiency despite the lower luminous efficacy compared to LPS.

B.8 Metal Halide

Metal halide lamps originally used quartz or arc tubes sealed with the arc gases and metal salts within it. This gave a good quality light with acceptable colour rendering for tasks where white light was required. The small arc tubes allowed good optical control, which increased system efficiency. The main problem with quartz metal halide was the deterioration of the light over time. Not only would there be a significant reduction in the amount of light produced, requiring over-lighting to compensate, but the colour of the light would drift from white towards the green or red tones. The lamp life was also limited at 8,000 to 12,000 hours requiring regular lamp changes to maintain a quality maintained installation.

The introduction of ceramic arc tubes for metal halide lamps improved the consistency of light output over time. Lumen depreciation and colour shift was reduced and lamp life was improved.

Recent developments of ceramic metal halide by Philips have produced the Cosmopolis lamp. This lamp has a shorter arc tube than HPS and metal halide, which allows further improvements in optical control. The lamps were designed to operate on electronic control gear, allowing 15 % efficiency improvements in the system efficiency. Since its launch, the lamp life has extended from four to six years as standard, reducing the maintenance costs and allowing the bulk lamp change, clean and electrical test maintenance regimes to be implemented, giving savings at the project level. The main disadvantage of the Cosmopolis system was the increased costs relative to HPS and metal halide solutions resulting from high lamp costs, new and expensive lamp holder designs and the cost of electronic gear. Cheaper solutions are now available commercially using standard Edison Screw (ES) lamp holders and electronic gear designed to mimic the operation of the Philips electronic control gear. These lower cost solutions provide access to some of the advantages of the full Cosmopolis system but the overall system efficiencies, lamp life and light quality are not necessarily equivalent.

B.9 LEDs

By 2009, improvements in the light output and luminous efficacy had enabled LEDs to be used for some exterior illumination, although the first significant installation of LED exterior lighting in UK was not until 2011. In 2013, the luminous efficacy gains in LEDs and the improvement in LED drivers and controls made this the best technical solution available for energy efficiency in most cases, but high capital costs prevented its widespread adoption. Since then, lantern costs have reduced and the technology is now being widely adopted throughout the UK. LED luminaires available at the time of writing have luminous efficacy ranging from 100 to 130 lm/W depending on drive current selected although it is expected that this will continue to improve as LED technology advances.

Funding sources

C.1 Public sector funding

C.1.1 Introduction

Local government spending is about a quarter of all public spending in the UK and is governed by the Local Authorities (Capital Finance and Accounting) (England) (Amendment) Regulations 2014. Local councils are funded by a combination of grants from central government, council tax and business rates. (In Northern Ireland, district councils still raise money through a domestic rate and a business rate). Local authorities also receive income from investments, council rents, sales and charges for services. Central government (or the devolved government in Scotland, Wales and Northern Ireland) provides specific and general grants to enable local authorities to deliver all the necessary services.

When considering local authority budgets with respect to the street lighting services the budget will be a mixture of both capital and revenue budgets. Capital budgets look to the costs of new works and improvement schemes whereas revenue budgets look to the operating costs of the street lighting asset, be they energy, maintenance or operational costs.

C.1.2 Capital budgets

Spending is classed as capital if the item or asset purchased has a life of more than one year. A full definition that has legal status is set out in the *Statement of Recommended Practice* published by the Chartered Institute of Public Finance and Accountancy. Examples of capital spend are:

- constructing and improving roads;
- replacing an age expired lighting installation; and,
- energy reduction programmes.

Money for capital programmes can be obtained through:

- borrowing (within limits based on prudence, affordability and sustainability);
- government grants and other external contributions;
- sale of assets, for example, property; and,
- contributions from our current year (revenue) budgets.

The operational costs for capital projects are met through the revenue budgets.

A local authority can also meet spending requirements by drawing on money saved in the past (reserves).

C.1.3 Capital programme, Prudential Code

The Local Government Act 2003 allows an individual local authority to borrow money to fund capital spending subject to the plans being prudent, affordable and sustainable in line with the Chartered Institute of Public Finance and Accountancy (CIPFA) prudential code for capital finance.

This code has brought about a change in focus, encouraging local authorities to consider whether projects meet the required objectives and allowing revenue and capital proposals to be considered, together within the budget process, i.e. permitting energy revenue savings to be used to pay the capital investment load.

C.2 Revenue budget

C.2.1 Funding considerations

In law, local authorities are not permitted to borrow money to fund revenue spending.

When considering any funding option it is important that a due diligence review has been undertaken and that the existing asset is known and understood so that any energy, carbon and service savings can be accurately determined. It is therefore vital that a local authority considering any funding programme should have a high level of confidence in its inventory of unmetered energy agreements, which may be taken as being when the inventory is over 97.5 % accurate.

If, for example, energy reduction funding is sought, then funders will each have their own requirements for application. Some focus purely on funding for energy saving equipment, light sources, control equipment, luminaires and the like whereas others will also seek funds for other assets such as lighting columns. Some may also look to the sustainability of the installation with regard to good asset management avoiding potential downstream issues.

C.2.2 Salix funding

Salix Finance Ltd. delivers interest-free loans to the public sector to improve their energy efficiency and reduce their carbon emissions. Salix has been working with the public sector in England, Scotland and Wales for over ten years and is delivering significant results. It has funded over 12,500 projects valued at over £300 million, saving taxpayers £80 million a year. Projects already supported are projected to save over £1.1 billion in energy costs over their lifetime.

Salix funding has the following set of compliance criteria:

- projects must be 'additional' – i.e. it would not have happened without this funding such as secured budgets are already in place or the project is required by legislation;
- the funding provided should have a simple payback of less than five years and in the case where this may not be possible, then the public sector body will bring their own or other additional funding to support the overall cost; and,
- the cost of CO_2 must be less than £100 per tonne over the lifetime of the project.

As it is funding for the public sector, then it is good to acknowledge that it is the public sector that needs to fully benefit from the resultant energy savings over the lifetime of the technology installed.

There is no maximum loan value and Salix has provided funding for projects exceeding £5 million. Projects are expected to be delivered and commissioned within 9 months, subject to National and European tender requirements. For large lighting projects that are planned over a number of years, Salix can work with clients to support their funding needs.

Salix is able to fund any elements which are required to allow the energy efficiency element of the project to be installed, for example controls and columns. However, as above, all equipment purchased must be within the compliance criteria.

In terms of support, Salix provides a Project Compliance Tool, which allows clients to easily check that their project applications meet the Salix compliance criteria for their respective loans programme. A completed compliance tool is required for each funding application made through Salix.

Applications can be made through a simple online process on the Salix website. They also assess the business cases, host regional meetings to share client knowledge and have examples of case studies and project knowledge slides to share with both existing and new clients.

For more information on Salix, please see http://www.salixfinance.co.uk/

C.2.3 Public Works Loan Board (PWLB)

Some public sector organisations, such as councils, have statutory powers to borrow and can access low cost borrowing through the Public Works Loan Board (PWLB). Interest rates are typically lower than commercially available loans, so public sector organisations that have access to PWLB will probably find this a cost effective route.

Most long-term council borrowing currently comes from the PWLB as it offers competitive interest rates and flexible terms.

C.2.4 Public funding schemes and competitions

Additional public sector funding may be found through UK and European funding schemes and competitions. See the DECC publication, *A guide to financing energy efficiency in the public sector*, for further detail.

C.2.5 Green Investment Bank funding

The Green Investment Bank (GIB) is 100 % owned by the UK Government and invests in projects which are both green and profitable. It offers funding to local authorities for energy efficiency projects, including low energy street lighting projects, which is made at commercial rates. It is able to invest in both on and off balance sheet structures and has developed a specific on balance sheet product for local authorities: the Green Loan.

GIB has worked with a number of local authorities across the UK in order to develop the Green Loan which is specifically designed to meet the needs and circumstances of local authorities.

The Green Loan is tailored to each project to which it is applied. Unlike Salix funding, it can be used to fund all related project spend, including columns, CMS and project development costs. It also offers local authorities the ability to:

- match the loan drawdowns to the expected spend profile of the installation of the equipment;
- fix the interest rate for the duration of the loan;
- defer debt repayments (capital, interest and fees) during the installation phase;
- sculpt debt service thereafter to match the forecast savings; and,
- choose the repayment period, anywhere up to 30 years, to match project needs.

Deferring debt payments during the installation phase and then matching payments thereafter to match forecast savings allows projects to be structured as a spend to save and they can be tailored to be cash flow positive in each period.

GIB work alongside local authorities to help tailor funding to specific projects and can also assist in developing the business case.

C.2.6 Private sector funding of local authority projects

Some private sector funders will lend directly to local authorities for on balance sheet solutions and some may also lend to Energy Service Companies (ESCOs) for an off balance sheet solution. These investors will invest at commercial rates.

Street lighting energy efficiency programmes hold an attraction to organisations keen to invest in green projects. It must be understood that these organisations lend on commercial terms. The approach taken may not just look at the primary energy and carbon savings but also look to consider secondary environmental effects and causes and good asset management is one of these considerations. They look to ensure suitable due diligence through the use of competent lighting professionals and a good inventory to use as a benchmark for savings and the basis of the business case.

Funding from these organisations is not limited to just the energy efficient equipment, as is the case with Salix funding, but can consider the rest of the infrastructure although payment is through the energy and carbon savings as well as reduced maintenance costs and the application of smart lighting such as trimming and adaptive lighting. Repayments are matched to energy savings.

These organisations do not have any prescribed delivery roll-out for the investment but from examples seen would perhaps look to a three-year implementation plan with a 20-year loan period. Interest may only be payable once the funding is drawn down for use and various profiles can be looked at to best suit the project. They recognise that payback periods may well be of the order of 6 to 16 years.

The requirements of a Salix bid provide some guidance on what would be required for any bid. This includes:

- having a high level of confidence in the inventory;
- establishing the performance requirements for the lighting sources;
- looking to the application of a CMS to allow trimming and adaptive lighting;
- full cost analysis;
- a phase programme approach;
- technical and financial modelling to support the predicted savings;
- procurement options;
- pilot studies; and,
- how the efficiencies will be measured and recorded.

C.3 Funding/incentives available to private sector organisations

C.3.1 Enhanced Capital Allowances scheme

The Enhanced Capital Allowance (ECA) scheme gives tax incentives for companies to install energy-efficient equipment including lighting. The ECA scheme allows 100 % of the capital sum to be written off against corporation tax in the first year after installation, thereby offsetting the initial capital cost of the higher quality lighting scheme. Detailed criteria have been published covering high-efficiency lighting units (luminaires), white LED luminaires and lighting controls.

Revised in 2013, the ECA scheme for white LED luminaires sets out minimum luminaire efficacy depending on the type of lighting system and on the use of lighting controls at:

- 60 luminaire lumens per circuit Watt for amenity, accent and display lighting regardless of the use of controls;
- 60 luminaire lumens per circuit Watt for general indoor lighting using downlights with dimmer and photocell control;
- 65 luminaire lumens per circuit Watt without dimmer and photocell control;
- 75 luminaire lumens per circuit Watt for general indoor lighting using uplights with dimmer and photocell control;
- 80 luminaire lumens per circuit Watt without dimmer and photocell control; and,
- 65 luminaire lumens per circuit Watt for exterior area and floodlighting luminaires regardless of the use of controls.

The luminaire efficacy values above must be met after 100 hours of continuous operation, and, in addition, all luminaires are required to provide at least 90 % of their initial light output after 6,000 hours of continuous operation. A power factor of at least 0.7 is also mandatory at all levels of light output, and a standby power of maximum 0.5 W is required for individual control gear integrated in electronic dimming or switching circuits.

The list of detailed criteria can be consulted online at: https://etl.decc.gov.uk/etl/site/etl/browse-etl/lighting.html

C.3.2 Green Public Procurement

Green Public Procurement (GPP) is a voluntary instrument that is aimed to help the EU economy to increase its resource and energy efficiency. Recommendations for public procurement have been published on a range of products including indoor lighting and street lighting. These cover the efficacy of lamps and the overall power consumption of the whole installation.

The GPP criteria for indoor lighting, published in 2012, require that LED lamps are classified Class A, with the exception of cases where a colour rendering index above 90 is required when LED lamps of Class B are accepted. Other requirements relevant to LED lighting cover lamp life (minimum life of 15,000 to 25,000 hours depending on lamp and control gear types), packaging (use of recycled materials) and maximum lighting power consumed in different building types.

LED lighting is also addressed by the GPP criteria for street lighting, revised in 2012, under the lighting design section where energy efficiency requirements are imposed for the whole lighting system based on the road class. Hence a lighting design calculation is required to verify compliance of LED products for each application.

The full lists of GPP criteria can be consulted online:

- for indoor lighting: http://ec.europa.eu/environment/gpp/pdf/criteria/indoor_lighting.pdf
- for street lighting: http://ec.europa.eu/environment/gpp/pdf/criteria/street_lighting.pdf

C.3.3 Energy Saving Trust Recommended

Energy Saving Trust Recommended is a certification and labelling scheme for most energy efficient domestic products available on the market. Its aim is to help consumers save on domestic energy consumption. Specific criteria were also introduced for performance, packaging and quality of LED lamps and luminaires, for which an independent third party test report must be produced by manufacturers as certification proof to demonstrate compliance. Luminaires accepting replacement LED lamps are not included in the scheme.

The current Energy Saving Trust Recommended criteria for LED lamps and LED luminaires are EST LED Requirements for Replacement Lamps and Modules Version 2.0 – 2010 and EST LED Requirements for Luminaires Version 3.0 – 2010, respectively.

The full list of criteria can be found on the Energy Saving Trust website at: http://www.energysavingtrust.org.uk/

Regulations, standards and guidance

D.1 Legislation

Artificial Optical Radiation Directive (2006/25/EC)

Clean Neighbourhoods and Environment Act 2005 Road Lighting and the Environment

Construction (Design and Management) Regulations 2015

Construction Products Regulations 2013

Control of Artificial Optical Radiation at Work Regulations 2010

Ecodesign Directive (2009/125/EC)

Electricity at Work Regulations 1989

Health and Safety at Work Act 1974

Low Voltage Directive (2006/95/EC)

Managing Health and Safety at Work Regulations 1999

Restriction of Hazardous Substances (RoHS) Directive (2002/95/EC)

Waste Electrical and Electronic Equipment (WEEE) Directive (2012/19/EU)

WEEE Regulations 2013

D.2 British and European Standards

BS EN 40-3-1:2013. Lighting columns. Design and verification. Specification for characteristic loads

BS EN 40-3-2:2013. Lighting columns. Design and verification. Verification by testing

BS EN 40-3-3:2013. Lighting columns. Design and verification. Verification by calculation

BS 5489-1:2013. Code of practice for the design of road lighting. Lighting of roads and public amenity areas

BS 5489-2:2003+A1:2008. Code of practice for the design of road lighting Part 2: Lighting of tunnels

BS 7671:2008+A3:2015. Requirements for Electrical Installations. IET Wiring Regulations. Seventeenth edition

BS EN 12193:2007. Light and lighting. Sports lighting

BS EN 12464-2:2014. Light and lighting. Lighting of work places. Outdoor work places

BS EN 12899-1:2007. Fixed, vertical road traffic signs. Fixed signs

BS EN 13201-2:2003. Road lighting. Performance requirements

BS EN 13201-3:2003. Road lighting. Calculation of performance

BS EN 13201-4:2003. Road lighting. Methods of measuring lighting performance

BS EN 13201-5 (in development). Road lighting. Energy performance indicators

BS EN 62471:2008. Photobiological safety of lamps and lamp systems

PD 6547:2004+A1:2009. Guidance on the use of BS EN 40-3-1 and BS EN 40-3-3

PD CEN/TR 13201-1:2014. Road lighting. Guidelines on selection of lighting classes

D.3 Institution of Lighting Professionals (ILP)

GN01: Guidance Notes for the Reduction of Obtrusive Light

GP02: Laser, Festival and Entertainment Lighting Code

GP03: Code of Practice for Electrical Safety in Highway Electrical Operations

GP09: Lighting the Environment - A Guide to Good Urban Lighting

GP10: Safety During the Installation and Removal of Lighting Columns and Similar Street Furniture in Proximity to High Voltage Overhead Lines

PLG02: The Application of Conflict Areas on the Highway

PLG03: Lighting for Subsidiary Roads: Using White Light Sources to Balance Energy Efficiency and Visual Amenity

PLG04: Guidance on Undertaking Environmental Lighting Impact Assessments

PLG05: The Brightness of Illuminated Advertisements

PLG06: Guidance on Installation and Maintenance of Seasonal Decorations and Lighting Column Attachments

PLG07: High Masts for Lighting and CCTV (2013 edition)

PLG08: Adaptive Lighting (not yet published)

TR12: Lighting of Pedestrian Crossings

TR22: Managing a Vital Asset: Lighting Supports

TR23: Lighting of Cycle Tracks

TR24: Practical Guide to the Development of a Public Lighting Policy for Local Authorities

TR25: Lighting for Traffic Calming Features

TR26: Painting of Lighting Columns

TR27: Code of Practice for Variable Lighting Levels for Highways

TR28: Measurement of Road Lighting Performance on Site

TR29: White Light

TR30: Guidance on the Implementation of Passively Safe Lighting Columns and Signposts

D.4 Society of Light and Lighting (SLL)

Guide to Limiting Obtrusive Light 2012

SLL Code for Lighting

SLL Lighting Guide 04: Sports Lighting

SLL Lighting Guide 06: The Outdoor Environment (SLL LG6)

SLL Lighting Handbook

D.5 Institution of Engineering and Technology (IET)

Code of Practice for Electrical Safety Management

Code of Practice for the Application of LED Lighting Systems

IET Guidance Notes to BS 7671

D.6 Countryside Commission

Countryside Commission/ Lighting in the Countryside: Towards good practice

D.7 Highways England, The Scottish Office Development Department, The Welsh Office and The Department of the Environment for Northern England

Design Manual for Roads and Bridges (DMRB):
http://www.standardsforhighways.co.uk/dmrb/

Volume 0 Part 1 (GD 02/08)
Quality management systems for highway design

Volume 2 Section 2 Part 1 (BD 26/04)
Design of lighting columns

Highway Electrical Registration Scheme (HERS) / National Highway Sector Scheme 8
Highway Electrical Registration Scheme (HERS)

Traffic Signs Regulations and General Directions (TSRGD)

D.8 International Commission on Illumination (CIE)

CIE S 004/E-2001: Colours of Light Signals

CIE 031-1976: Glare and Uniformity in Road Lighting Installations

CIE 032-1977: Lighting in Situations Requiring Special Treatment (in Road Lighting)

CIE 039.2-1983: Recommendations for Surface Colours for Visual Signalling, 2nd ed.

CIE 047-1979: Road Lighting for Wet Conditions

CIE 061-1984: Tunnel Entrance Lighting: A Survey of Fundamentals for Determining the Luminance in the Threshold Zone

CIE 066-1984: Road Surfaces and Lighting (Joint Technical Report CIE/PIARC)

CIE 088:2004 (2nd edition): Guide for the Lighting of Road Tunnels and Underpasses

CIE 094-1993: Guide for Floodlighting

CIE 115:2010 (2nd edition): Lighting of Roads for Motor and Pedestrian Traffic

CIE 126-1997: Guidelines for Minimizing Sky Glow

CIE 129-1998: Guide for Lighting Exterior Work Areas

CIE 132-1999: Design Methods for Lighting of Roads

CIE 136-2000: Guide to the Lighting of Urban Areas

CIE 140-2000: Road Lighting Calculations

CIE 143-2001: International Recommendations for Colour Vision Requirements for Transport

CIE 144:2001: Road Surface and Road Marking Reflection Characteristics

CIE 150:2003: Guide on the Limitation of the Effects of Obtrusive Light from Outdoor Lighting Installations

CIE 154:2003: The Maintenance of Outdoor Lighting Systems

CIE 189:2010: Calculation of Tunnel Lighting Quality Criteria

CIE 193:2010: Emergency Lighting in Road Tunnels

CIE 194:2011: On Site Measurement of the Photometric Properties of Road and Tunnel Lighting

CIE 206:2014: The Effect of Spectral Power Distribution on Lighting for Urban and Pedestrian Areas

D.9 Useful links

British Standards Institution (BSI): www.bsigroup.com

Campaign to Protect Rural England (CPRE): www.cpre.org.uk

Chartered Institute of Public Finance and Accountancy (CIPFA): www.cipfa.org

Elexon: www.elexon.co.uk

EU FP7 Humble Lamppost project: ec.europa.eu/eip/smartcities

Green Investment Bank (GIB): www.greeninvestmentbank.com/

Health and Safety Executive (HSE): www.hse.gov.uk

Highways Electrical Association: www.thehea.org.uk

Highways England: www.gov.uk/government/organisations/highways-england

Illuminating Engineering Society (IES): www.ies.org

Institute of Asset Management: (IAM): theiam.org

Institution of Engineering and Technology (IET): www.theiet.org

Institution of Lighting Professionals (ILP): www.theilp.org.uk

International Commission on Illumination (CIE): www.cie.co.at

International Electrotechnical Commission (IEC): www.iec.ch

Society of Light and Lighting (SLL): www.cibse.org/society-of-light-and-lighting

UK Roads Liaison Group (UKLG): www.ukroadsliaisongroup.org

D.10 Resources

BEAMA Guide to Surge Protection Devices www.beama.org.uk

Cochrane Database of Systematic Reviews http://community.cochrane.org/cochrane-reviews; including:

Beyer FR, Ker K. Street lighting for preventing road traffic injuries. Cochrane Database of Systematic Reviews 2009, Issue 1. Art. No.: CD004728

HEA / ILP Guide to the Intelligent Management of Public Lighting Life Safety Engineering Systems: http://downloads.thehea.org.uk/index.php/hea-technical

Highways Maintenance Efficiency Programme www.highwaysefficiency.org.uk/

HMEP UKRLG Highway Infrastructure Asset Management Guidance - www.ukroadsliaisongroup.org/en/utilities/document-summary.cfm?docid=5C49F48E-1CE0-477F-933ACBFA169AF8CB

Local Partnerships Efficient Lighting - http://localpartnerships.org.uk/our-work/growth/efficient-lighting

Managing Unmetered Energy Street Lighting Inventories (ILP): www.theilp.org.uk/documents/unmetered-electricity/

Prioritising Investment in Public Lighting-A framework for developing a Street Lighting Value Management Model (ILP): https://www.theilp.org.uk/documents/prioritising-investment-in-public-lighting/

Secured by Design www.securedbydesign.com/pdfs/110107_LightingAgainstCrime.pdf

Scottish Futures Trust (SFT) Street Lighting Toolkit www.scottishfuturestrust.org.uk/publications/street-lighting-toolkit/

UK Roads Liaison Group Code of Practice 'Well-Lit Highways.' www.ukroadsliaisongroup.org/en/utilities/document-summary.cfm?docid=220C1896-5D20-4A54-B010156913910E69

Glossary

NOTE: A number of lighting terms are discussed in 2.2 Lighting design principles and terminology.

Abrupt Failure Value
The percentage of LED light sources or luminaires of the same type that no longer give any light at the Median Useful Life. For example, the Rated Median Life may be 50 000 hours for a depreciation to 70 % of the initial light output. At this time the number of abrupt failures could be 6 %. This would be written as L70 = 50 000 AFV = 6 %.

Adaptive lighting
The ability to change the lighting levels using any of a group of controls that includes, Central Management Systems (CMS), part-night lighting, stand alone dimming and Multi Level Static Dimming.

Ballast
Component that provides the required voltage to start a discharge lamp (for example, fluorescent or high intensity discharge lamps) and then limits and regulates the amount of current supplied to the lamp during operation; can also accept an input signal to set the output brightness level.

Beam angle
The angle between points on opposite sides of a beam at which the luminous intensity is half the maximum value.

Blue light hazard
The potential for retinal damage due to high intensity exposure of the retina to light at wavelengths between 400 nm and 500 nm.

Chromaticity coordinate values
Chromaticity coordinates are a set of three numbers (x, y and z in the International Commission on Illumination (CIE) system) that give a unique colour. A chromaticity diagram shows the colours that correspond to each set of coordinates. MacAdam ellipses refer to the region on a chromaticity diagram which contains all colours that are indistinguishable to the average human eye, from the colour at the centre of the ellipse. The contour of the ellipse therefore represents the just noticeable differences of chromaticity. MacAdam ellipses are often scaled up to a larger size of 3x, 5x or 7x the original. This is indicated as a 3-step, 5-step or 7-step MacAdam ellipse. Initial variations in colour, or colour shift with time, can be shown by the size of ellipse needed to contain these positions. The initial and maintained chromaticity coordinates are measured for the maintained value at 25 % of rated life up to a maximum of 6 000 hours.

Colour gamut
A measure of how colourful objects appear under a given light source. It is obtained from the area in a chromaticity diagram of the polygon formed by joining the chromaticity coordinates of reference colour samples when viewed under the light source (see Chromaticity coordinate values). Colour rendering (Ra), colour rendering index (CRI). An indicator of how accurately colours can be distinguished under different light sources. The colour rendering index (measured in Ra) compares the ability of different lights to render colours accurately. This measures the ability of a light source to render colours naturally, without distorting the hues seen under a full spectrum radiator (such as

daylight). With some LED technology having a narrow spectrum, however, the CRI index is not in all circumstances giving a fair representation of the colour appearance, therefore new definitions and methods for measuring the representation of colours are currently under development by CIE. The colour rendering index (CRI) ranges from 0 to 100. Natural daylight is rated at Ra = 100.

Colour spectrum
The distribution of colours produced when light is dispersed by a prism.

Colour temperature (kelvin, K)
Correlated Colour Temperature (CCT) The colour temperature provides an indication of the light colour and is expressed in kelvin (K). Lamps are generally rated between 2700 K (warm) and 6500 K (cool/daylight). The higher the colour temperature, the cooler the perception of the white light becomes.

Compact fluorescent lighting (CFL)
A type of fluorescent lamp designed to replace incandescent lighting.

Contrast
The relationship between the luminance of an object and its background.

Curfew Time
Time during which lighting is dimmed or switched off, usually to save energy or reduce obtrusive light.

Cylindrical illuminance
Total flux falling on the curved surface of a (vertical) cylinder divided by its area. This is used as a measure of how much light falls on people's faces.

Efficacy
A measure of light output against energy consumption measured in lumens per watt.

Electrical power (watts)
A watt (W) is a unit of measurement for power equal to one joule of energy per second. The amount of electrical power required or consumed by a fixture or appliance is known as its wattage (voltage x amperage = wattage).

Electronic control gear
Most commonly described as the driver. A unit that is located between the electrical power supply and one or more LED modules in order to provide the LED module(s) with an appropriate voltage or current. It may consist of one or more separate components, and may include additional functionality, such as to act on control signals sent to it, means for dimming, power factor correction, radio interference suppression and feedback.

Energy (watt-hours)
Energy is the amount of power that is used over a period of time. The most common unit used for energy is a kilowatt-hour (kW h). For example, a 100 W lamp operated for 10 hours uses 1 000 watt-hours, or 1 kWh.

Environmental zones
Areas with specific requirements for lighting control e.g. residential areas versus town centres. Local Planning Authorities may specify the following environmental zones for exterior lighting control within their Development Plans ranging from E1 intrinsically dark landscapes such as Areas of Outstanding Natural Beauty to E4 high district brightness areas with high levels of night-time activity.

Flicker
The impression of the rapid and repeated variation with time of a lamp's brightness or (less usually) colour.

Flux
The sum of all lumens emitted by a light source.

Gateway
A device for interfacing between two networks that use different protocols, which may use different data speeds, data volumes and data volumes.

Gear
see Electronic control gear.

Glare
The uncomfortable brightness of a light source or lit object against a darker background, which results in dazzling the observer, or may cause nuisance. Glare can be caused by either direct or reflected light. Reflected glare is the result of bright reflections from light, polished or glossy surfaces. Direct glare occurs when the light travels directly from the source to the eye.

Heat sink
The part of the luminaire that conducts the heat away from the LED.

High pressure sodium lamp
An HID lamp whose light is produced by radiation from high pressure sodium vapour (and usually a small amount of mercury).

Horizontal illuminance (E, Eh)
Illuminance incident on the horizontal surface. Unit: lux (lx) $= lm/m^2$.

Illuminance (lux (lx))
The amount of light falling on a surface of unit area. The unit of illuminance is the lux, equal to one lumen per square metre.

Incandescent lighting
Light produced when a filament is heated to incandescence using an electric current.

Isolux diagram
A diagram showing lines joining points of equal illuminance.

LED
Also known as a light-emitting diode, an LED is a solid state semiconductor that emits light. LEDs are used in a variety of lighting applications and now consume less power than many conventional lamps sources for the same delivered quantity of light.

LED module
A unit supplied as a light source. In addition to one or more LEDs it can contain further components(for example, optical, mechanical, electrical, and electronic components, but excluding the electronic control gear).

Lens
A transmissive glass or plastic optical system that diffuses, concentrates or redirects light.

Lighting controls
A broad category of technologies that in general control the lighting based on various inputs. Lighting controls include dimmers, absence/presence detectors, photo/multi-sensors, relays, remote-controls timers and switches.

Light engine
A subsystem used to generate light, which typically includes a lamp module, optics and projection lens.

Light output ratio (LOR)
Ratio of the total light emitted by a luminaire to the total light output of the lamp(s) it contains measured at standard operating conditions.

Light source
Any device serving as a source of illumination.

Light spill
see Obtrusive light.

Light trespass
see Obtrusive light.

Longitudinal uniformity
Ratio of the lowest to highest road surface luminance along the centre of a driving lane.

Lumen (lm)
A unit of measurement that expresses the total quantity of light given off by a source, regardless of direction. One lumen is equal to the amount of light that one candle emits over one square foot of surface that is exactly one foot away from the flame.

Lumen depreciation
The decrease in lumen output that occurs as a lamp or light source is operated over a period of time.

Luminaire
A complete lighting fixture consisting of one or more lamps or light sources, along with the socket connections and other parts that hold the lamp in place and protect it, wiring that connects the light source to a power source, and a reflector/lens or other optical system that helps direct and distribute the light.

Luminance (cd/m²)
The luminous flux emitted by a surface in a given direction divided by the area of the surface. It is a measure of the brightness of a surface.

Luminance ratio
A ratio used to characterise absolute variation in surface brightness for a defined field-of-view.

Luminance uniformity
Ratio of minimum luminance to average luminance.

Luminous efficacy (lm/W)
The total luminous flux emitted by the light source divided by the lamp wattage; expressed in lumens per watt (lm/W).

MacAdam ellipse
see Chromaticity coordinate values.

Maintained illuminance (lux)
Value below which the average illuminance on the specified surface is not allowed to fall. The maintained illuminance is specified at the end of the maintenance cycle, taking into consideration the maintenance factor. It is one of the main specification elements for the lighting designer. In the various lighting standards the maintained illuminance is specified for various areas/activities.

Maintenance factor
Correction factor used in lighting design to compensate for the rate of lumen depreciation, caused by lamp ageing (lumen depreciation and lamp failure) and dirt accumulation (luminaire and environment). It determines the maintenance cycle needed to ensure that illuminance does not fall below the maintained value.

Median Useful Life (Lx) parameter
The time for 50 % of a batch of products to fade to x % of its original light output (when driven under constant current drive conditions). Usually, an L70 value is quoted, meaning the time to drop to 70 % of the initial light output (deemed appropriate for functional lighting applications as the value beyond which the light output becomes unsuitable for purpose).

Mesopic vision
Vision when the eye is adapted to conditions between daytime (photopic) and night-time (scotopic) luminance levels. Most street lighting falls into this range.

Obtrusive light
Any light emission that has a negative impact on the surroundings. This can take the form of sky glow, light spill, light trespass or glare.

Optic
The components of a luminaire such as reflectors, refractors, protectors which make up the light emitting section.

Overall uniformity
Ratio of the lowest to highest road surface luminance.

Photoperiod
Response of fauna and flora to the period of daylight.

Polar curve
Graph showing the luminous intensity of a light source at different angles.

Rated input power
Shows the amount of energy consumed by a luminaire (expressed in watts).

Rated lamp life
The life value assigned to a particular lamp type or lighting system. This is commonly a statistically determined estimate of average or median operational life. For certain lamp types other criteria than failure to light can be used; for example, the life can be based on the average time until the lamp type produces a given fraction of initial luminous flux.

Reflectance
Ratio of the luminous flux reflected from a surface to the luminous flux incident on it.

RGB (red, green and blue)
The three primary colours of light, which can be mixed to create white light. Also refers to the colour model for displays and monitors, where combinations of illuminated red, green and blue pixels are used to create a wide variety of colours.

Sky glow
see Obtrusive light.

Source intensity

This is the brightness of the source of the luminaires and applies to each source in the potentially obtrusive direction, outside of the area being lit.

Thermal management

The ability to control the temperature (heat) of the device junctions in packaged LEDs, often through the use of heat sinks. Junction heat can negatively impact the performance of LED lighting, including output, colour and lifetime.

Thermal resistance

The measure of a material's resistance to heat flow. In packaged LEDs, thermal resistance is used as an indirect method of determining LED junction temperature.

Vertical illuminance (lx)

Illuminance incident on the vertical surface.

INDEX